FORT OF DAULATABAD

FORT OF DAULATABAD

DR. DULARI QURESHI

BKP

BHARATIYA KALA PRAKASHAN

DELHI (INDIA)

First Published : 2004

***Published by* C.P. Gautam**
BHARATIYA KALA PRAKASHAN
3421-A, IInd Floor, Narang Colony, Tri Nagar,
Delhi-110035. Phone : 27393083, 30967505 (O), 27522129 (R)

ISBN : 81-8090-072-X

Laser Type Setting by
A-ONE GRAPHICS
JD-18C, IInd Floor, Pitampura, Delhi-110053
Phone : 55640278, Mob. : 9811167357

Printed By :
BDH Printer at Salasar Imaging Systems
Delhi, Ph. 31045804, 30930620

DEDICATED

With Deepest Love & Respect.
To my Guru, Dr. R. S. Morwanchikar,
A Renowned Scholar and historian, in
the field of Art, Culture, Heritage and
Water Conservation.
Former Director & Head of History
Department, and Tourism Administration.

CONTENTS

PREFACE

In the world of art and architectural scholarship I owe my initial training to Late Dr.Ramesh Shankar Gupte, my father, who has authored several books on Ajanta, Ellora and a number of other cave temples of India. As authority on iconographical studies, he presented me with a deeper perception of cave temples in analyzing minute details of each and every aspect which otherwise would escape one's attention. After my father's untimely death my learning process was passed on to Dr. R.S. Morwanchikar, my guide and guru who assisted me in a more scientific process of studying and examining a monument as it is closely related with the study of art and architecture, that was when I developed a practice of examining, scanning and observing every detail not only concerning the monument but around the monument and its surrounding.

In the present book I have tried to accomplish a pleasant assignment which my publisher had thrust on me as there was no single book on Daulatabad fort while the other forts of India were much in focus due to several books available on them in the market. I have long felt that our traditional method of writing on art and architecture does injustice to the present status of several ancient and medieval structures especially if we admit only limited aspects of descriptive analysis of a monument. I have slightly deflected by including not only the historical and architectural aspect but also rather included the present status of each and every structure and what is required for its long standing survival.

I have however refrained from including the social, economical, literary or religeous aspects of historical study but preferred to restrict myself to a brief history and architectural study of the structures in and around the fort. However, I have also undertaken several educational tours for the students of my department of Tourism, I had consistently been drawn to the shortcomings and weaknesses of our tourist destinations, reason why I decided to examine the surroundings of the fort and understand the 'why's of the deterioration and erosion of the structure the shabby conditions of its environments which are responsible in reducing the pleasant experiences of the tourist while moving in and around the fort.

I do not need to be told how absurd this enterprise is, or how immodest is its very conception. An attempt to relate both architecture and tourism might not really be appreciated by those who would rather like to limit architectural studies within a narrow compass of history, sociology or political relationships, but I have made an effort to deviate and relate both, in tourism perspective.

Nevertheless I have dreamed that despite the many errors inevitable in this kind of work it may be of some use to those agencies who are making a herculean effort not only to preserve and conserve our heritage for posterity but also to transform its periphery for a greater tourist experience.

Indian historical remains scattered in every nook and corner of the country portray a picture of the wonderful past but because we have inherited hundreds of such historical remains sometimes too many concentrated within the precincts of one destination, they have been criminally neglected and the consequences are there for all to witness. Our historical heritage is our wealth and it not only earns us foreign currency but we must understand that without heritage we will become rootless. Karan Grover, an architect, town planner and conservator feels that we are what because of our heritage. While promoting tourism the planners and developers have to be very careful in assessing what the site can take and the average time one could spend there.

7th Sept. 2004
Aurangabad

Dr. Dulari Qureshi

ACKNOWLEDGEMENT

The appellation Aurangabad has always been synonymous with Ajanta and Ellora while Daulatabad fort, which is in closer proximity to Aurangabad, has been conveniently unallied. Very few books, articles or even guidebooks are available on the fort. In fact some officials from the Archaeological Survey of India, requested me to write a guidebook on the forts, as very little information is available for the tourists or is of a sporadic nature.

The chapter on Architecture was supported by secondary data and party on field surveys while the chapter on environmental impacts of tourism development in the fort was mostly supported by field surveys and is a part of the research project sponsored by the Vice-Chancellor Shri K.P. Sonawane, Dr. Babasaheb Ambedkar Marathwada University and Dr. V.S. Lomti Director, Board of College and University Development and Dr. Surase, the Registrar of the Dr. Baba Saheb Ambedkar Marathwada University. It is a collection of information compiled together from sample survey through interviews as well as observation and use of old photographs, articles and maps.

Several officials who assisted me with valuable information through interviews, proved very useful. Shri Amar Rathod, the Regional Manager, Department of Maharashtra Tourism Development Corporation apprised me with the latest efforts of the corporation in making an effort at the regression of the shops so that more space availability will be a boon and the surrounding ugliness will recede. Mr. Yadav, the Manager, Department of Tourism, Govt. of India, Aurangabad, also in a telephonic conversation complained of the unresponsive attitude of shopkeepers thus spoiling the architectural beauty of the fort.

Mr. Khalid Ahmed is not only a history lover but an avid photographer who has frequently rambled around the fort and is well equipped with detailed knowledge about its surrounding monuments. His huge collection of photographs of Daulatabad fort and its surrounding monuments have offered me an insight into some of the most minutia problems around the fort which otherwise would have escaped my attention.

I am also grateful to Dr. Mrs. Mohal the Librarian, Dr. Mrs. Lomte the present Librarian, Mrs. Saraswati More and Mrs. Shaista Khan for their valuable assistance. I must thank Shri Bagul for his special assistance in tracing art books. My special thanks however are due to Mr. Sumar who is adept at procuring the most untraceable books.

My mother late Dr. Nalini Gupte was the greatest source of inspiration. My sister and brother Dr. Mrs. Nita Patwardhan and Mr. Sachin Gupte have encouraged me to write the book. My two sweet sons Shakir Qureshi and Saber Qureshi have tolerated my absence from home for long periods.

However my principal source of inspiration and encouragement is my dear husband Mr .Rafat Saeed Qureshi who helped me in all my efforts. He accompanied me patiently during the number of times to Daulatabad, photographed and procured photographs and maps for the book.

My father-in-law late Shri Abdul Hameed Qureshi had given me several inputs. I must not forget to thank the I/C Director Dr. S.A. Somavanshi of the Dept. of Tourism Administration who has always encouraged me with all my research work.

I must thank my publisher Mr. C.P. Gautam, due to whose persuasion this book was completed. It was on his suggestion that the book has been written.

Last but not least my sincere thanks are due to Mr. Manoj Rayalwar, the steno-typist of our department who patiently and meticulously typed my book.

Aurangabad
9th Sept. 2004

Dr. Dulari Qureshi

LIST OF PLATES

LIST OF FIGURES

INTRODUCTION

As the day draws to the close the dark silhouette of the Daulatabad fort presents an enignatic, foreboding picture of the past. The huge sprawling hillock with innumerable, monumental walls rising beneath, arresting the entry of the strongest armies, a clear deterrent against invasion is one of the most unique examples of military stronghold of the medieval period.

Several dynasties have contributed in creating fortifications and making it stronger through several decades. Early writers, travellers,scholars, historians have visited the fort and recorded their impressions of their visit.

Ibn-e-Battuta was one of the earliest travellers who is said to have accompanied the sultan Muhammed Tughlaq in 1340 to Daulatabad has documented his imprints in a book called Safarnamah which was later translated. The present book Ibne Battuta's Travels in Asia and Africa has been written in English by George Routledge.[1]

Ibn-e-Battuta writes: From Ujjayni we travelled to Daulatabad, Aurangabad, the enormous city which rivals Delhi, the capital in importance and in the spaciousness of its planning. It is divided into three sections, one is Daulatabad proper and is reserved for the sultan and his troops, the second is Kataka and the third is the citadel which is unequalled in strength and is called Duwaygir (Deogiri).

"At Daulatabad resides the great Qutlag Khan, the Sultan's tutor, who is governor of the fort and the Sultan's representative. This province is well populated and is wholly under his authority and that of his lieutenants. The fortress of Deogiri mentioned above is a rock situated in a plain, the rock has been excavated and a castle built on it's summit. It is reached by a ladder made up of leather which is taken up at might. In its dungeons there are huge rats, bigger than cats- infact cats run away from them and cannot defend themselves against them, so they can be captured only by employing ruses. The inhabitants of Daulatabad belong to the tribe of Marhata (Marathas) whose women have been endowed with special beauty particularly in their noses and eye brows. The inhabitants of this town are merchants dealing principally in jewels and their wealth is enormous. In Daulatabad there is an exceedingly fine and spacious bazar for singers and singing girls containing numerous shops each of which has a door leading to the house of the proprietors.

In the centre of the bazar there is a large carpeted and decorated pavilion in which the musician sits every thursday after the afternoon prayers. In the same bazar there are mosques for the prayer service."[2]

It is extremely difficult to pinpoint the precise location of the bazar, or the pavilions and the mosques. Whether they were in the interior of the main fortifications or in the outskirts is difficult to state.

Ibne Battuta graphically described the transfer from Delhi to Daulatabad when Muhammad Tughlaq had ordered all the inhabitants of Delhi to migrate to Daulatabad. To assure that every individual had left the capital the Sultan, according to Battuta one night climbed the roof of his palace and looked out over Delhi, where there was neither fire nor smoke nor lamp and said that now his mind was tranquil and his wrath appeased.[3]

In the India Gazetteer the Deogiri fort is described as follows. The fortress is built upon a conical rock, scarped from a height of 150 Ft. from the base. The hill upon which it stands rises almost perpendicularly from the plain to a height of about 600 ft. It was first captured by the Muhammedan in 1294 and Sultan Muhammad Tughlaq recognized its importance as a base for military operations in Southern India, renamed it Daulatabad and concieved the idea of making it his capital.[4]

After Deogiri the name of the fort was not Daulatabad but Qubbatul-Islam and coins were struck with this name of the mint engraved on them. Three sections of Daulatabad are mentioned, the city of Daulatabad, Deogiri and Katak. Sometimes it has also been referred to as Dharagir, Dharakhera and was situated at the top of the hill.[5]

The official chronicler of Shah Jahan's Court Ahmad Lahori and a historian and author of the great work Padshah-Nama[6] described the fort thus. "This lofty fortress, the ancient names of which were Deogiri and Dharagir and which is now known as Daulatabad is a mass of rock which raises its hand to heaven. The rock has been scraped throughout its circumference, which measures 5000 league yard[7], to depth which ensures the retention of water[8] in the ditch at the foot of the escarpment. The escarpment is so smooth that neither an ant nor a snake could scale it. Its height is 140 cubits[9] and around its base a ditch forty cubits in width and thirty in depth has been dug in the solid rock. Through the centre of the bill a dark spiral passage like the ascent of a minar, which it is impossible to traverse, even in day light,without a lamp and has been cut, and the steps in this passage are cut out of the rock. It is closed at the foot of the hill by an iron gate and after passing through this and ascending, the passage one enters the citadel.At the head of the passage is a large grating of iron which is shut down in case of necessity and when a fire is lighted upon it the ascent of the spiral passage becomes impossible owing to the intense heat. The ordinary means of reducing portions such as mines, covered ways, batteries etc. are useless against this strong fortress. This passage still exists and is the only work, the attribution of Muhammad Tughlaq is still doubtful, because Ibn-e-Battuta who visited the fort in 1342 or thereabout records that access to the citadel was then gained by means of a leather ladder[10].

Ahmad Lahori writes further, "From the foregoing account it may be surmised that though some of its defences were devised by the Yadavas and other by the Bahmanis and their successors, some also were the work of the Muhammad Bin Tughlaq. These were carried at the time when he transferred his capital from Delhi to Daulatabad and constitutes one of the most striking example of fortification known to the medieval world. The fortress is an an isolated height of 700 ft.forming an outlier to the Balaghat range that bounds it on the north and east. The hill is cone-shaped and the city of the Daulatabad was situated to the east and south of it, but the greater portion is now in ruins, only a few small huts are surviving. Round about this almost impregnable acropolis which possesses its own perennial springs of water was a highly complex system of fortification designed to protect the city. The outer wall had a perimeter of 2¾ miles and between it and the acropolis were three inner walls, each loop holed and battlemented and each furnished with fortified gateways, out works and bastions all so disposed that with the help of salient and re-entrant angles the maximum of five could be directed against an assailant.

From the Baradari, a small gateway to the rear of the pavilion gives access to the bastion on the summit which was used to fly the flag by the successive rulers. The hill here is 30ft. in circumference and has a wall with small bastions at the angles. The principal bastion carries a gun- The pathway from the right of this bastion leads to a cavern which was the residence of a Hindu ascetic by name Janardhan Swami. The Swami flourished in the 16th Cent. and was the Prime Minister and the chief adviser to the Governor of Daulatabad. He was the preceptor of Eknath Maharaj of Paithan. He wrote several works but unfortunately few of his writings have been preserved. On Margashirsha Vadya, a fair commences in honour of the Swami. It lasts for three days and on the last day is held Kala which is attended by 15000 persons.[11]

Another discription of Daulatabad is available by both Thevenot and Tavernier who visited Daulatabad around 1666 A.D. Thevenot writes, "This town was the capital of Balaghat before it was conquered by the Mughals. It belonged than to the Dakhan and was a place of great trade. It is walled round with free stone and has battlements and towers mounted with cannons. But through the walls and towers be good, yet this is not the thing that makes it accounted the strongest place belonging to the Mughals. It is a hill of an oval figure, which the town encompasses on all sides, strongly fortified and having a wall of a natural smooth rock, Thanenvirous is at the bottom, with a good citadel at the top whereon the king's palace stands. Besides the citadel, three other parts of the fort are called bar-kot, mar-kot and third kala-kot. It took 2.1/2 hours to reach Daulatabad from Aurangabad."[12]

According to Asgar Bilgrami the distance between Daulatabad and Golconda was a journey of 15 to 16 days.[13]

In early studies by Henry George Briggs on Daulatabad is typical of the period. He refers to the fort as a place of "great celebrity, eight miles to the westward of Aurangabad. Its original Hindu name was Deogur and it is conjectured to be the Tagara of Ptolemy. It was reduced by the Mohemmedans just before Warangal and appears at that time to have been the capital of the old Mahrathha country, of the extent of which, as of Telingana.[14]

Early in the 14th Cent. Emperor, Tughlaq Shah removed the seat of government from Delhi to Daulatabad.The fortress is a stupendous work, but its strength is an object rather curious than useful. The principal fortress is on a rock about 600 ft. high, the edge of which has been scarped away so as to leave the principle all round 1500 yds. in length and varying from 200 to 50 to 300 ft. in perpendicular height.

At the foot of the scarping there is a ditch near 55ft wide and 45ft deep. The entrance is from the inner gates that are built one within the other upon the plain and lies over a narrow bridge across the ditch into a small door, from which a shaft ascends spirally through the heart of the rock and is set by an iron trap rock in the centre of the work above. The shaft is about, 300 yds. long about 7 ft. high and 10ft. wide. The lower fort consists of stone walls flanked by bastion. They have little strength and even the upper works, which are confined and have not much cover, might perhaps be bombarded[15].

One of the most detailed books with pictures and line drawings covering all the provinces of Deccan is the Glimpses of the Nizam's Dominions written by Claude Campbell in 1898. The pictures are extremely interesting due to the period when not much construction was visible around. The experience of witnessing these unique historical creations was far more satisfying than they are in the 21st Century.

According to Mr. Campbell's description Daulatabad was nine miles north west of the town of Aurangabad and had a population of 1500. The fortress was built upon a conical shaped rock scarped from a height of 150 ft. from the base. The bill upon which the fort stands rises almost perpendicularly from the plains to a height of about 600 ft. The fort has been provided with a counter scrap gallery and a complete system of countermines. The outer wall is 2¾ miles in circumference, and between the wall and the base of the upper part there are three inner lines of fortification. The outer wall formerly enclosed the city, which was built round the base of the fortress, but this is now for the most part one mass of ruins. The village which is now all that remains of the ancient city of Deogiri, contains but a few huts, which are principally occupied by his Highness the Nizam's Police force stationed in the fort for the place is now a state prison. On the summit of the rock is a flagstaff and a cannon. There are no buildings of import.[16]

A few yards further is the only entrance to the upper part. Consisting of a narrow stone bridge crossing a moat about 30 ft wide. This was built by Salar Jung I, but previously the moat was crossed by three or four planks.[17]

A steep ascent from the shrine leads to the handsome pavilion near the summit. This is built of stone and is evidently of Hindu construction. Report says that it was the residence of Hindu princes of Deogiri, and it was probably used by them as a summer resort, and also by the emperor Shah Jahan and son Aurangzeb.[18]

Another interesting sketch of the fort is offered by Shri Kamalapur in his book The Deccan Forts. "Perhaps the most complex and intricate of all the forts in the Deccan, the capital of Maharashtra for over a century (1187-1294) and for a short period the capital of Islamic India under the 14th Cent. Sultanate of Delhi, Daultabad or former Deogiri is a hill cum land fort situated 9 miles north west of Aurangabad on the main road from that city to Ellora of the famous rock cut temple (this road in fact actually runs through a part of the fortification). The fort is on an isolated cone shaped hill of rock rising abruptly from the plain about 700' and situated by the ancient ghats or roadway running to Ellora. A minimum of artificial defence would be necessary on a hill so strongly fortified by nature to make it deserve the name of a stronghold"[19]

The fortification constitutes three concentric lines of walls, with innumerable bastions, culminating in the citadel throughout the wall stands on the inner side of the rock cut ditch and is in addition to the bastion further provided with formidable, gateway with elaborate local defences. The sides of the hill have artificially scarped along the entire circuit of the mass to form on all sides a vertical rock cut form varying from 150' to 200' in height with a great square section water filled moat at the bottom also dug 50' ft into solid rock. The outermost wall encloses an irregular oval area about 1.1/2 miles by 3/4 miles linking up with the citadel hill at north and south, is 2 miles and 6 furlongs in circumference. There were originally seven gateways of which only two called Mecca and Rauza are in use now. The entrance led by a flight of steps constitute of two gateways and has a semi-circular tower and a Nagarkhana. The area inside upto the next line of fortification forms the town proper has been visited by Ibn-e-Battuta in 1340.[20]

Discussions : The account of Ibn-e-Battuta is one of the earliest descriptions since he visited the fort during its period of wealth and prosperity. The Magnitude of the city during the Tughlaq period can well be imagined as Ibn-e-Battuta compared it with Delhi in importance and spaciousness. It must also have embraced a large area as it was divided into three sections. Ibn-e-Battuta mentions a castle built on the summit of the fort. The castle on the summit generally known as Baradari is believed by most scholars like Raunaq Ali to have been built on the instructions of Shah Jahan. As one witnesses the architectural details, the four centric arches on the interior as well as the exterior, the typical decorative embellishment of stucco plaster and the geometrical designs all point to its Mughal ownership. There is a definite possibility that in the ancient period of the Yadavas the lower structure must have been constructed which is visible even today as the building stands on an octagonal stone structure supported by brackets on the lower side. This too has been renovated. The upper portion of the Baradari is definitely a structure of the Mughal period. Shah Jahan must have demolished the original building and resuscitated a new structure in the Mughal architectural style. The second description of singers performing in the public places seems to have been a common form of entertaintment.

Secondly the name of Deogiri was changed first into Qubbatu'l Islam as is stated by Haroon Khan Sherwani in his book. 'The Bahamanis of the Deccan', and later it was called as Daulatabad by Muhammad Tughlaq but the citadel according to several available references continued to be called as Deogiri for a long time. The different name like Dharagir, Dharakhera are in all probability misinterpretations or misconstruction of the original word. Such distortions are common throughout history. Since people of different races, countries like Turkist, Persians, Iranians, Mughals came from different regions such distortions are to be expected.

Thirdly when Ahmad Lahori, Shah Jahan's court chronicler came to Daulatabad he vividly described the famous dark passage inside the fort. But according to Lahori the ownership of this passage to Muhammad Tughlaq is doubtful as the access to the citadel in 1340 by Ibn-e-Battuta was gained by means of a leather ladder. But according to strong references and architectural details the dark passage was there and its owenership goes not even to the Tughlaq period but the Yadava period. The entrance of the dark passage has a doorway decorated on the top with temple motifs and has a recessed doorway. The pillars too have a base, a shaft and a capital all of which is plain.

The fourth point is the height of the fort which is given as 600 ft. in the India Gazetter. While Ahmad Lahorl gives it as 700 Ft, while Henry Briggs states that the height of the principal fortress is 600ft., Campbell gives the same height. However the confirmed height of the principal fort is 700ft.

Henry George Briggs fifthly has conjectured Daulatabad to be the Tagara of ptolemy. Tagara is in no way connected with Daulatabad but is modern Ter which is connected with Ujjani on a caravan route.[21] The road came to Bhogvardhana (Bhokardan) then to Rajatadated (Aurangabad) and from here proceeded to Pratishthana (Paithan) and on to Tagara (Ter) as stated by Schoff W.H. in Periplus of the Erythrean Sea.[22]

Sixthly during the Mughal period strong protective iron gates were fixed, gateway through the dark passage, small gateway to the rear of the pavilion, several such contivances made the entrance of enemies near impossible. Again guns were mounted after every interval, while on the lower portion its fortification walls had battlement and towers also mounted with guns and cannons.

Ibn-e-Battuta had stated that the fort was reached by a leather ladder. It was probably the moat which was crossed by a leather ladder as a permanent fixture would have allowed access to the enemy. Though a number of these scholars and historians visited the fort none of them have bothered to describe the buildings, their ownership and their architectural details. Only a passing reference circumscribed within a single page or less has been permitted for the accounts of Daulatabad Fort.

It is also clear from these accounts that during the Nizam's (Asaf Jah-I) period Daulatabad was deserted and probably the hub of all activities shifted to Aurangabad. As Campbell records that the fort during Nizam's period was used as a state prison and his police force is stationed there.

Unlike most forts of India with definite and continous ownership of one particular personality or a single dynasty Daulatabad fort has a different history. Though initially excavated and built by the Yadava dynasty in the 12th Century A.D. the present fortifications, expansions and building activities within a radius of a few kilometers is the ingenuity and accomplishment of several diverse dynasties extending through a span of at least five centuries and recurrent invasions and occupations. Due to its strategic location and position, its strong protective defenses, (famously called as an impregnable fort) its possession was craved by most powerful dynasties ruling between 12th and 17th century A.D. its ownership became a matter of pride and prestige and haughty arrogance. No wonder no particular dynasty could retain the fort for a long period.

One reason why it has become so difficult to definitely pinpoint the architectural creations of fortification walls, palaces, masjids, temples to one particular dynasty. Though comparative architectural parallels of each period have assisted in establishing the style to a particular dynasty.

Again very limited number of scholars and historians have done comprehensive research work on Daulatabad except for Dr. Riaz Ansari. After Dr. Ansari work no academician has ever taken the trouble to cover all the aspects like socio-religious-economic-historical-architectural-literary and technical review of the fort.

The present book "Daultabad fort" is also restricted to the architectural creations in and around Daulatabad fort and the environmental impacts of tourism development on all these creations in the present era. This research study is partially directed towards the modern problems of unplanned development, illegal encroachments, no maintenance of the numerous buildings, by A.S.I., complete indifference of authorities towards the gradual destruction of the buildings. There is little sense of responsibility or accountability on the part of various dhabas, restaurants, souvenir shops, fruit stall owners, or any sense of hygiene or cleanliness located around the fort. The transporters park the vehicles in close vicinity of the fortification walls creating continuous injury to the walls.

The first chapter is a sketchy outline of the historical details of the successive dynasties ruling in Daulatabad during various periods. It also includes information about the rulers, their period and achievements. This chapter also highlights the special contribution of important rulers and the political upheavals of their times. Though it is in no sense an exhaustive history, an effort has been made to discuss in some details the role of important rulers.

The second chapter is an all inclusive description of the architectural creation of all the rulers. The fortification walls of the fort which form an important component of the chapter as also the various darwazas, moat, bridges, caves, temples, masjids, step wells, hamam, tanks, palaces, have been described in minutia-details. Particulars of doors, ceilings, domes, arches, pillars have been delineated in microscopic details. Chand Minar the land mark of the Indian minarets has been specially focused as it is the second highest minar after Qutub Minar in Delhi, and is reputed for its glory and grandeur, built by Ala-ud-din Bahamani.

The third chapter highlights the water supply scheme at Daulatabad probably Amilan the brain child of the water scheme has been discussed at length. As most of Malik Ambar's water system rested on the technique of under-ground lime pipes, sections for silting and filtering chambers and air towers have been given special attention.

The fourth chapter covers monuments around the fort. Daultabad fort's precints spread and covered Khuldabad, Kagzipura upto Ellora Caves, it also inclued Maliwada, Lasur, Nizampur and a lot of the surrounding area. This is clear from the vestiges of monuments all around the fort like the Sarai at Nizampur, the Eidgaah at Maliwada, the Pariyaon Ka Talab near Shoolibhanjan and several such good pieces of architecture and historical remains. These structures do not find even a passing mention in most of the books. Thus most of the information is based on primary data collected through observation and interviews with old families.

However while going back in history the research work has carefully attempted to review the present status of all these monuments, their condition,their environmental problems and hazards of development for tourism growth. The work is a meticulous survey of all the surroundings and its impacts on the valuable heritage of Daulatabad. The environmental impacts of development, unplanned growth is gradually eroding the aesthetic beauty of the fort. It appears more congested and shabby. The periphery of the fort requires a complete face-lifting so that it satisfies the tourist to the maximum. These environmental problems have been discussed at length in the fifth chapter of the book.

REFERENCES :

1. Battuta Ibn.Travels in Asia and Africa (1325-54) Published by George Routledge & Sons. London 1929, p. 227.

2. Ibid., p. 228.
3. Ibid., p. 185.
4. Ibid., p. 363.
5. Sherwani Haroon Khan—The Bahamanis of the Deccan Krishnavas International, Hyderabad 1953, p. 39.
6. Srivastava Ashirbadi Lal- Medieval Indian Culture, Published by Shiva Lal Agarwala & Co. Agra 1964, p. 128.
7. League is a nautic measure 1/20th of a degree 3 geographical miles.
8. Aurangabad District Gazetteer, Govt. of Maharashtra, Aurangabad District, Revised Edition 1977, p. 966
9. Cubit is an old measure, the length of the arm from the elbow to the tip of the middle finger, from 18 to 22 inches.
10. Aurangabad District Gazetteer, p. 967.
11. Ibid., p. 971.
12. Aurangabad District Gazetteer, p. 966
13. Bilgrami Asgar S.A., The Landmarks of the Deccan, p. 163.
14. Briggs Henry George: The Nizam His History and Relations with the British Government Vol. II. London. Published by I "Spottiswoods and Co., 1861, p. 246.
15. Ibid., p. 247.
16. Campbell Claude: Glimpses of the Nizam's Dominion Published by C.B.Burrow. Bombay and London. 1898, p. 387.
17. Ibid., p. 388
18. Ibid., p. 389.
19. Kamlapur J.N. The Deccan Forts. Popular Book Published Bombay 1960, p. 106.
20. Ibid., p. 107.
21. Gupte Ramesh Shankar. The Iconography of the Buddhist Sculpture (Caves of Ellora) Published by Marathwada University, Aurangabad 1964, p. 17.

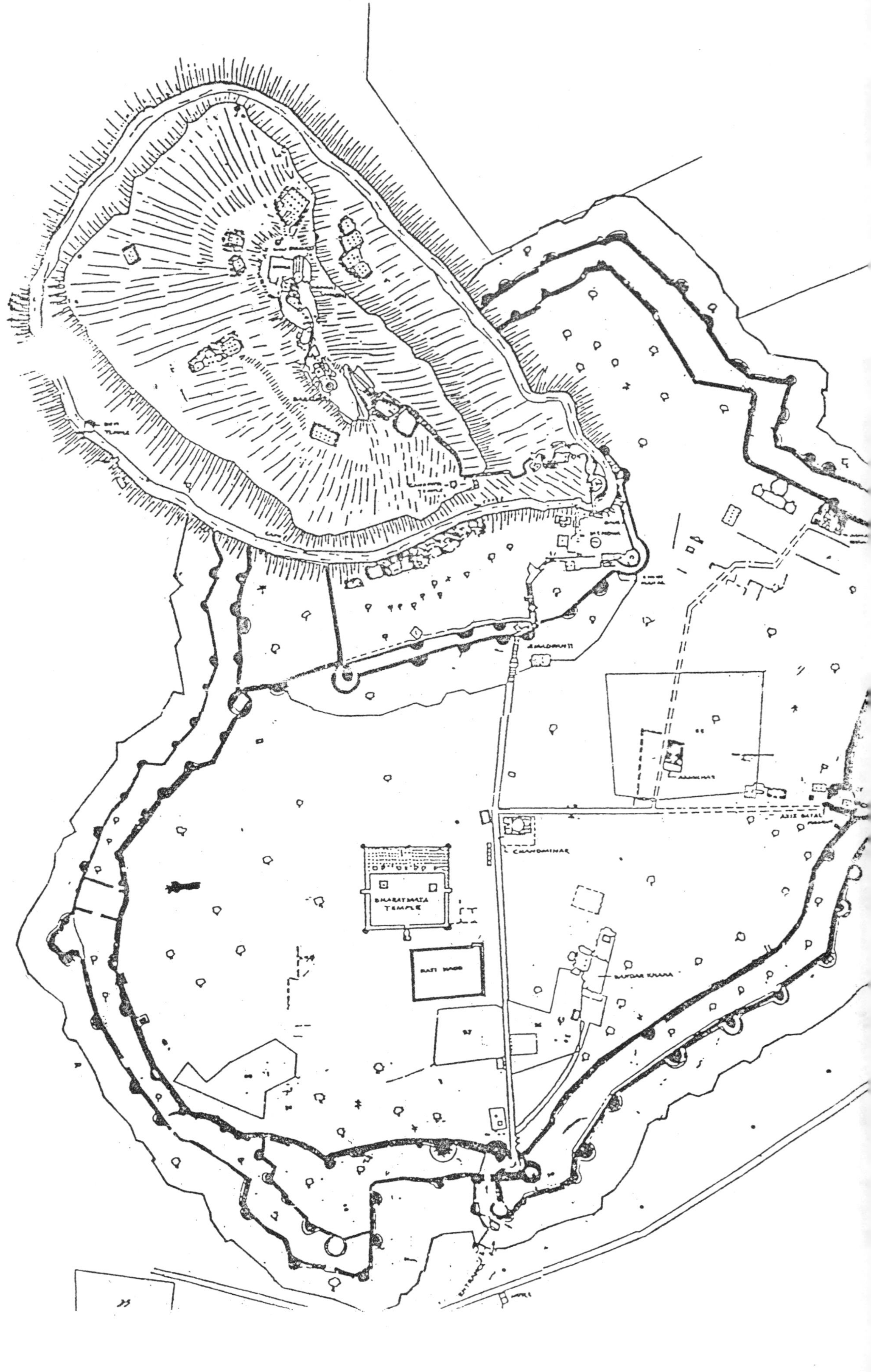
CHANDMINAR
BHARATMATA
TEMPLE

THE SITE PLAN OF DAULATABAD FORT

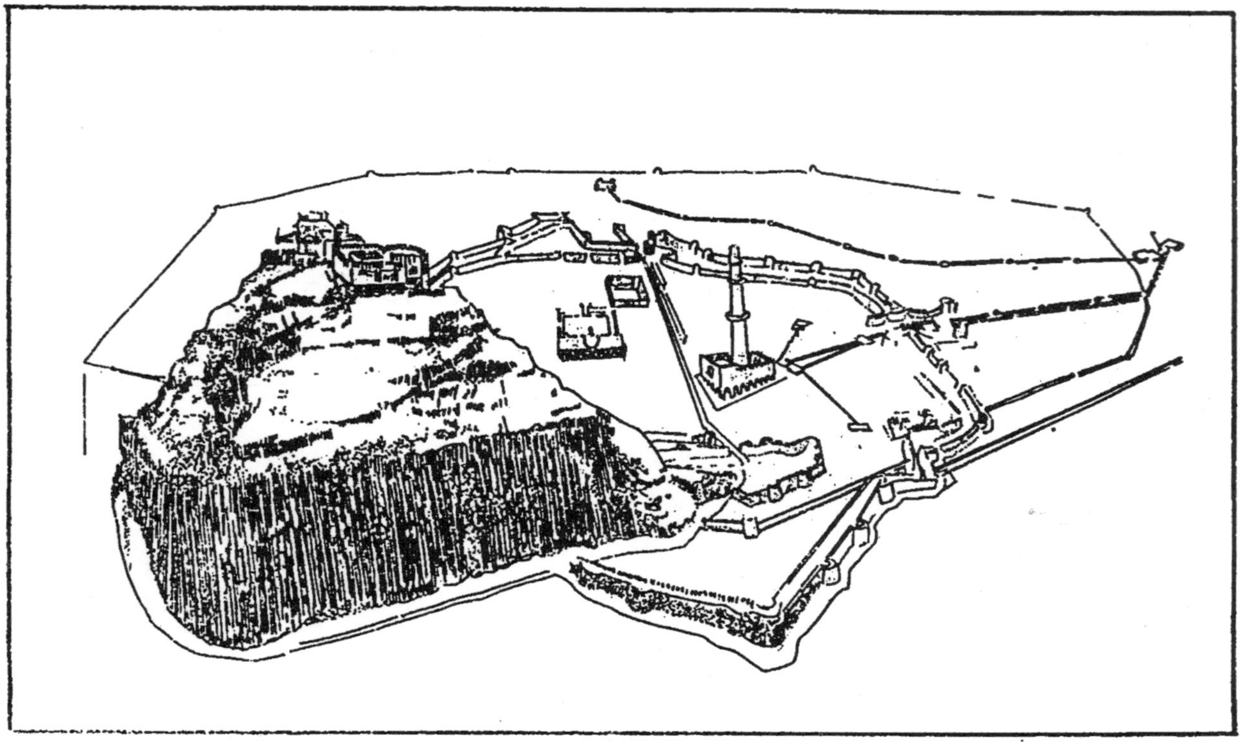

Fig. IIA. General view of Daulatabad Fort.

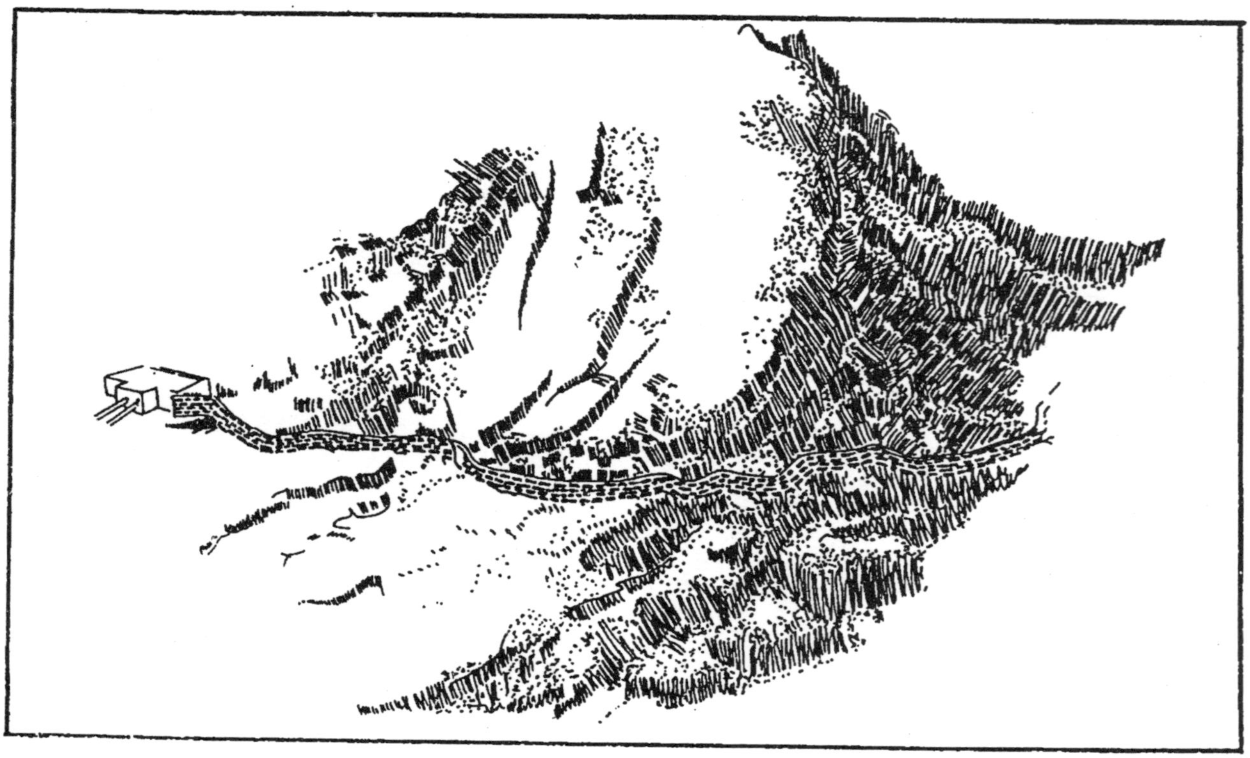

Fig. IIB. General view of Daulatabad Fort.

CHAPTER - I

A BRIEF HISTORY OF DAULATABAD

For a close scrutiny of the period and its glorious history a quick historical perspective of the various dynasties who ruled the fort would not only be interesting but also assist in a better understanding of its significance as an immortal monuments which ought to be conserved and preserved more scientifically and seriously for posterity. Amongst the few well maintained monuments, Daulatabad can also boast, but there is a gradual atavism in its conservationary efforts and alarming changes are taking place in marring it's external and internal environment. A travel through its past, the important dynasties who ruled and developed the fort into an invincible stronghold with the most ingenious, resourceful, adroit, protective and defensive devices will help understand its unique qualities.

Yadava Dynasty :

The founder of the Yadava dynasty according to reliable sources was Bhillama V (1187-1200) though the dynasty originates from a ruler named Seuna Chandra[i]. But for our present study we are not concerned with the earlier rulers as our focus is on Devagiri fort. So the ruler who has been directly responsible for its excavation was definitely Bhillama V who founded the city of Devagiri and having got himself crowned made that city his capital[1]. The date of ascending the Yadava throne is given as 1185-86 in the Yadavas and New Times by O.P. Verma[2] , while some books give the dating as 1187[3]. Probably Bhillama V ascended the throne somewhere between 1185 and 1187. The Yadavas according to V.A. Smith in the Oxford History of India[4] were considered as the feudatories nobles of the Chalukya kingdom. Bhillama V was the first imperial ruler of the Yadava dynasty to launch upon military conquests[5].

Bhillama V was succeeded by his son Jaitugi I, also known as Jaitrapal who launched a sudden hurricane attack against the Kakatiyas [6]. Jaitugi was succeeded by Singhana II, the most powerful ruler of the dynasty, who defeated the Chalukyas, the Kakatiyas and the Hoyasalas of Dwarasamudra. Singhna was succeeded by Krishna his grandson in (1246—61 A.D.) Krishna invaded Malva and defeated the joint enemies of the Chalukyas and Hoyasalas. Krishna was succeeded by his younger brother Mahadeva (1261-1271 A.D.). North Konkan was invaded and probably connected.

Ramchandra or Ramdeva assumed imperial power in (1271-1311 A.D.). The hostilities between Yadavas and Malva, Chalukyas and Hoyasalas continued. Yadava King was victorious against the Chalukyas, Malva and Hoyasalas. In fact the Yadavas had reached the zenith of their power in 1292 A.D. when the most unpredictable events followed which abruptly eclipsed the strength and power of these Deccan rulers who had subdued most of their contemporaries in the Deccan and south. In 1292 A.D. Ala-ud-din Khilji, nephew of the ruling Sultan of Delhi Jalaluddin Firoz Shah Khilji, appeared out of the blue and caught Ramadeva unaware at Lasura 12 miles from Devagiri.[7] Ramchandra took refuge inside the fort but finally he surrendered due to insufficient food provisions inside the fort. Ala-ud-din carried a huge ransom in the form of gold, silver, jewelery, horses, elephants and camels. Ramdeva's son Shankaradeva who was away returned and he confronted the returning army of Ala-ud-din. Shankara was defeated and Ala-ud-din obtained huge wealth from him.

From this time onwards the powerful prosperous Devagiri was caught up in the power hungry grips of the sultan Ala-ud-din who was crowned in 1296. Ramchandra submitted and sent revenue failing which Ala-ud-din sent Malik Kafur. After Ramchandra's death in 1311 his son Shankara refused to pay the annual tribute and in 1313 Malik Kafur re-appeared at Deogiri. This was the fourth expedition of Malik Kafur to Deogiri in which Shankara was defeated and killed. Meanwhile Ala-ud-din died in 1316. At Deogiri Harapal the son-in-law of Ramadeva declared his independence. The successor of Ala-ud-din, Qutb-ud-din Mubarak Khilji started for Deccan in 1317[8]. Harapal was defeated and tortured to death. This ended the Yadava dynasty, completely vanquished and replaced by Muslim officers. Mubarak built a great mosque out of the demolished Jain temple.

Muhammad Tughluq (1325 to 1351 A.D.):

Though the next dynasty was established by Ghayas-ud-din Tughluq who did not involve himself in Deogiri affair. It was the more daring son of Ghayas-ud-din, Muhammad Tughluq who was attracted to Deogiri due to its strategic position, central situation and not constantly threatened by Mangol raids[9]. Ibn-e-Battuta, the chronicler of Tughluq dynasty claims that Muhammad Tughluq ordered all the inhabitants of Delhi to shift to Deogiri.[10] The transfer of all its inhabitants also included skilled craftsmen and artisans to build numerous structures at Deogiri.[11]

His most notable contribution at Deogiri or Daulatabad was the introduction of token currency, wherein brass and copper coinage was minted.[12] To this day at Daulatabad copper and brass coins can still be discovered. The capital at Daulatabad was established in 1326 and probably lasted upto 1334 A.D. when Muhammad Tughlaq realized his folly and returned to Delhi. Daulatabad was now a deserted fort, standing in it's monumental glory waiting for a new master.

Bahamani Dynasty :

It was said that in 1347 Hasan Gangu Bahamani, under the title of Abul Muzaffar Ala-ud-din Bahman Shah[13] established the Bahamani kingdom. Ala-ud-dins' capital was Gulbarga though during his life time Daulatabad too formed a part of his vast territories.

The Bahamani rulers were responsible for establishing some safety measures and re-strengthen the fort of Daulatabad by adding the fortification walls called the Kalakota and double walls of Mahakota with formidable gates and strong bastions.[14] Besides this some public structures and especially the Chand Minar was a creation of Ala-ud-din Bahamani.

Nizamshah of Ahmednagar :

The Bahamanis during the later period to a large extent was dependent on its powerful intelligent prime-minister, who was put to death by Muhammad Shah III, at the instigation of his ministers[15] and brought about the ruin of the Bahamani dynasty. After the downfall of the Bahamanis by 1480 the large empire broke up into five sultanates of Deccan. We are only concerned with Nizam Shahi dynasty of Ahmednagar under whom Daulatabad fort was also included. In the early period of its rule Daulatabad was not a very important fort as Nizamshahi concentration was on Ahmednagar and its surroundings. Daulatabad was captured by Malik Ambar the commander-in-chief of the Nizam Shah dynasty in 1499 A.D. which helped him to consolidate his dominions.[16]

Emergence of Mughal rulers created havoc in Ahmednagar and its precincts. Gradual invasions of the Mughal emperors weakened the Nizam Shah's dynasty. Akbar since 1591 sent his ambassadors to the Sultanate. In 1600 Ahmednagar was stormed and part of it came under the Mughals. It was during Jahangir's reign that Malik Ambar, commander-in-chief of Nizam Shah's forces since 1604 started dominating the Deccan politics. It was said that in warfare, in command, in sound judgement, and in administration, he had no equal.[17] Ahmednagar could not be conquered till Malik Ambar was alive. It was Malik Ambar who shifted the royal Nizam Shahi family to Daulatabad as the attacks of the Mughals became more and more ferocious. But under Shah Jahan the Mughals besieged the fortress of Daulatabad in 1631.[18] The incapable son of Malik Ambar, Fateh Khan had placed Nizam-ul-Mulk's son, Husain Shah, a boy of ten on the throne when the Mughals besieged the fortress. Fateh Khan surrendered the fort after taking a bribe of ten and a half lacs of Rupees[19] from the Mughals. Ahmednagar was annexed to the Mughal empire in 1633 A.D. and the nominal king Husain Shah, was consigned to life long imprisonment in the fort of Gwalior. Daulatabad thus passed into the Mughal hands during the period of Shah Jahan.

Mughals at Daulatabad :

Daulatabad was now used by the Mughals under Shaha Jahan and later

Aurangzeb as a strong base against the sultans of Bijapur and Golconda. In 1636 Shahjahan arrived in Daulatabad to recommence Mughal operations against Bijapur and Golconda.[20] Shahjahan was at Daulatabad for a few months, after which he went to Golconda and returned to Delhi around 1638-39. At the same time Shahjahan in order to gain stronger control of the Deccan placed his son Aurangzeb as a governor twice (1636—44, 1652—57) during which period he spent a lot of his time in continuous warfare with Bijapur and Golconda. Though Aurangabad was the capital of Deccan he also spent a lot of his time at Daulatabad fort. During his Governorship Daulatabad was captured by Mahabat Khan. During the absence of the Mughal kings the fort was under the charge of a custodian as is clear from the studies on Maratha History written by Madhavrao Pagadi.[21] Probably Daulatabad also started being used to imprison miscreants. A letter in this work written to the custodian of Aurangabad that a certain Khan Mohammed should be paraded in the streets and handed over to the custody of the commandant of the fort of Daulatabad.[22]

During Aurangzeb's tenure he often stayed at Daulatabad especially during his last 25 years which he mostly spent in the Deccan. He either stayed at Aurangabad, his official capital or Daulatabad.[23]

Nizam-ul-Mulk :

After Aurangzeb's death the Mughal empire started disintegrating. Aurangzeb's sons succeeded one after another. Due to continuous succession it became difficult to control the Deccan. Though for the first few years the domination of Mughals continued, by 1722—24 the Nizam-ul-Mulk, the wazir of Deccan virtually established an independent kingdom in the Deccan. While the Mughals found themselves in a defensive position against powerful invaders like Nadir Shah and Ahmad Shah Abdali an Afghan Chief, the Nizam Asaf Jah I continued to consolidate his position in the Deccan. The Nizam has been highly praised as an efficient administrator. His sons and grandsons continued to rule in the Deccan. But it was during the Mughal reign that Daulatabad gradually lost its glory. It has also been rumoured that the Nizam in fact was responsible for fleecing Daulatabad of its original glory. The Nizam mostly concentrated on Aurangabad initially and later on shifted his attention to Hyderabad. Though Daulatabad continued under commanders and killedars who were the caretakers and custodians of the fort, it lost its former glory.

Marathas :

The Peshwa Balaji Bajirao had won laurels in furthering the Maratha cause. Under him the Marathas were prospering, flourishing and expanding their wings. The Peshwa's cousin Sadasiv Rao inflicted a defeat on Nizam Ali at Udgir in 1760 A.D. A treaty was concluded by the latter with Nizam Ali by which the Marathas got the whole province of Bijapur, nearly the whole of Aurangabad and a portion of Bidar, together with some forts including the famous fortress of Daulatabad.

Unfortunately for the Peshwa this state of affairs did not last for a long time as the British gradually aserted themselves. The Nizam later by 1780 won back Daulatabad from the Marathas.

REFERENCES:

1. Bhandarkar R.G. Early History of the Deccan. Publisher Sushil Gupta Calcutta 1957, p. 117.
2. Verma O.P. The Yadavas And Their Times Publisher Dr. M.B. Deopujari, Nagpur 1970, p. 49.
3. Morwanchikar R.S. Deogiri—Daulatabad An Archaeological Overview. Publisher Rajwade Sanshodhan Mandal, Dhule 1993, p. 3.
4. Smith A. Vincent. The Oxford History of India, Oxford University Press, 1958, p. 219.
5. Verma O.P. The Yadavas And Their Times, p. 70
6. Ibid., p. 148
7. Sinha and Banerjee, History of India, Published in Calcutta, p. 246
8. Ibid., p. 264
9. Rushbrook W., History of India, p. 149
10. Sinha and Banerjee, p. 269
11. Srivastava Ashirbadi Lal—Medieval Indian Culture, p. 151.
12. Rushbrook Willian History of India, p. 151
13. Sinha and Banerjee, p. 299
14. Morwanchikar, p. 6
15. Majumdar, Ray Chaudhari and Datta Macmillian and Co. London, 1948, p. 361
16. Ibid, p. 364
17. Ibid, p. 467
18. Ibid, p. 475
19. Ibid, p. 475
20. Srivastava Ashirbadi Lal, Medieval Indian Culture, p. 308
21. Agadi Madhavrao—Studies in Maratha History Vol. II, Shivaji University Publications, Kolhapur, 1971
22. Ibid, p. 8
23. Smith A. Vincent, p. 387

CHAPTER - II

ARCHITECTURAL COMPOSITIONS WITHIN THE FORT

The architectural creations within the fort of Daulatabad leaves one amazed and confused. With so many successive and varied dynasties leaving their mark and impressions the fort offers the most complicated secret passages, hideouts, every self preservative defense in terms of voluminous fortification walls, with bastions, battlement, towers and fabricans, a wide moat, dark passages, ditches. Subterranean passages that cut through the citadel perplexing and bewildering the enemy. Besides the several fortification walls there are other distinctive buildings within the fort which could be classified as palaces, temples, mosques, hamams, minaret, pavilions, layers of habitation, streets. Walls and step wells. With all creations handed down to posterity the authorship of these structures has been difficult to trace and to this day remains a challenge to both the historian and the archaeologists. Though certain similarities dissimilarities have been brought to the forefront to draw certain conclusions these comparisons are definite.

FORTIFICATION WALLS OF THE FORT

The fort which according to the Indian ancient texts falls under the classification of Misra durg or mixed fort as it combines the qualities of Giridurg (mountain fort), Deva durg (God's fort) and Bhumi durg (land fort).[1]

The fortification walls were built at different periods by Yadavas, Khilji, Tughlaq, Bahamani, Nizam Shahi and Malik Ambar and bears a marked resemblance to medieval European forts in the disposition of projecting towers, covered passages and bastions.[2] The walls consist of labyrinth like arrangements of the underground secret passage with safety arrangements, in the style of buildings which give the bastions, despite their grim appearances, a touch of beauty by insertions of carved stones and dainty architectural details at various places.[3]

The strengths of Daulatabad Fort :

The fort has been considered as the most powerful, puissant fort that offered robust resistance to the enemies. An invincible fort, one of the best preserved forts in comparison to other forts in the vicinity of that same period. Its sound and sturdy quality exists due to the powerful wall about three miles in perimeter[4] and defended by a moat and glacis that runs round the hill at its foot and between this outer curtain and the citadel, also known as Dharagir and Duwagir.[5] The fort has been made more invincible due to the fact that it has been scarped all round vertically for a height of 150 feet with sides so smooth as to render escalade quite impractical.[6] The only approach to the citadel is across by a ditch or moat around a narrow, steep and tortuous tunnel with chambers for guard opening off at intervals.[7] Its basic design and character are Hindu as its originaters are the Yadavas who laid its foundations and in a lot of its characteristics it resembles the forts of Rajasthan. The smooth and lofty scarping round the base of the fort relates to a much earlier period for it is in the tradition of the skilled labours performed by the workmen at the Ellora excavation, only a few miles away. Additions have definitely been made by the Khiljis, Tughlaqs, Bahamanis and other successive dynasties, but the strong, huge work has been laid down by the Yadavas.

Fortified Walls :

Emergence of several different dynasties, invasion of Muslim rulers from the north, no recognized boundaries encouraged surprise attacks and invasion creating large scale insecurities for all rulers. Hence the necessity and major need was defense and fortification walls to a certain extent did provide security. Daulatabad fort and its land formation around was irregular and hence the wall constructed round the fort is irregular unlike other proximate forts.

The most interesting characteristic of this wall is that like Delhi or Aurangabad it is not single but is surrounded by several walls like Agra fort. According to Cambridge history of India the outer wall has a perimeter of 2¾ miles and between it and the acropolis are three inner walls each battlemented and each furnished with fortified gateways, outworks and bastions all so disposed that with the help of salient and re-entrance angle the maximum of fire could be directed against an assailant, in addition to which the outer wall is surrounded by a moat and glacis in much the same manner as medieval fortress of Europe. Though the Bahamins were largely responsible for the military engineering, genius, their successors were also responsible for introducing radical changes in military architecture due to the initiation of artillery.[8] Mahur and Naldurg were built on precipitous rocks and relied on artificial defence while other remarkable forts close to Daulatabad are Bidar and Parenda.[9]

Ambar Kot :

Ambar Kot was the exterior most wall surrounding the fort and is still popularly known as Ambar Kot. Built in mid-seventeenth century it is said to have been built by Malik Ambar, the Abyssinion commander of Nizam Shah of Ahmednagar and hence the name Ambar. It is a protective wall against the powerful Mughals who were contemporaries of Malik Ambar.[10] This fortification wall, enclosing, an irregular oval, is about 1.5 miles by 2¾ miles[11] linking up with the citadel at the north and south and having 9 gateways[12] and the present road from Aurangabad to Daulatabad and Ellora runs through it from south to north.[13] The outer two entrances used in the wall are Mecca gate and Rauza or Delhi gate. Mecca gate is towards the west and Delhi gate facing the road to Khultabad.

The outermost wall of the fort consists of 45 bastions massively built with their connecting curtain walls and nine principal gates with their flanking bastions and towers.[14] The walls have been built in stone and mortar locked up with a good breadth of rammed earth between the outer and inner casings. The height of the wall varies from 30 feet to 40 feet and it has lockhold battlements with circular towers at the angles on which once an occasional piece of antiquated ordnance is visible.[15] The bastions at irregular intervals are generally semi-circular or octagonal. Wall walks is a regular feature and are constructed from bastion to bastion, over gates and this is protected by high battlemented wall. Two different versions of the circumferences of the outer wall is found one by captain Seely and the second Raunaq Ali. According to the first opinion, the circumference of outermost wall is said to be five thousand yards and the thickness of the walls at the foundation is 15 feet and the height is 48 feet. The space within is divided into nine fortification walls separated by one another towards the center.[16] Raunaq Ali[17] says that the outer wall is three miles in parameter.

The wall has 45 bastions, nine principal gates with few postures. The names of the principal gates of Ambar Kot from south to north are as follows (Plate I).[18]

1. Fatehabad Darwaza
2. Nizam Shahi Darwaza
3. Paithan Darwaza
4. Kirti Darwaza
5. Naroti Darwaza
6. Kabdi Darwaza
7. Katgar Darwaza
8. Delhi or Rauza Darwaza
9. Khas Darwaza

All the gates are protected by flanking bastion generally semi-circular or octagonal in shapes. Most gates are double.

1. **Fatehabad Darwaza :** This is a village 5 km. north west of the fort and hence derives its name from it. The gateway comprises two arches the height of

which is 18 feet, while the total height including the parapet is 25 feet. On either side of the gateways are chambers 7 feet above ground level. The ceiling of the porch as well as the chambers is dome shaped. Steps near the inner arch on either side lead to the ramparts behind the battlements. Bastions are absent. Figures of lion on either side of the gateway.

2. **Nizam Shahi Darwaza :** This gate faces toward the east and probably derives its name from the village named Nizampur about 5 km. from the fort. This gateway too has two pointed arches. The height of the gate including parapet is about 30 feet. The porch has vaulted domed ceiling and has two chambers on either side. Either side circular bastions are constructed.

3. **Paithan Darwaza :** This gate faces towards the south east of the famous historical town of Paithan. The earliest gate as is evident from its architecture it has been so well strengthened that it is safest against enemies. The gate is huge in shape and has chamber with flanking balconies built in the pillar and lintel style.[19] The voluminous size is significant of its medieval importance as the main entrance gate to the fort. The height of gate including parapet is 30 feet. The ceiling is again dome shaped. There are figures of peacock on the brackets. However the upper roof of the Paithan gate is flat. Its facade is supported by battlements with machicolations. The upper portion of the bastion also has battlement and machicolation.

4. **Khirki Darwaza :** Since the gate faces toward the city of Aurangabad it is called as Khirki gate. The gateway comprises of two pointed arches.[20] The dimensions of inner and outer arches are uniform. The span of the porch is about 15 feet and the height upto the parapet is about 30 feet. There are chambers on either side just 2.5 feet above ground level. The ceiling of both porch and chambers are dome shaped. The facade is decorated with elephants with two men seated above. The gate has a bastion 15 feet away from the gateway built massively with parapets and machicolations. The front portion of the gate above the arch is a panel decorated with design purely of Hindu period. [21] Both the sides of the arch have the effigies of elephants standing on platform. The figures are carved in stone.

   The distance between Paithan gate and Khirki gate measures three furlongs and the plan is mostly a zigzag plan.

5. **Naroti Darwaza :** The Naroti gate comparatively is very small in plan and presently is closed. The height of the front wall of this gateway which also includes the parapet wall is 35 feet. The gateway comprises of two arches both uniform; the span being 12 feet. The gateway is flanked by bastions, semi-circular in shape.

6. **Kabdi Darwaza :** Interestingly the height of each gate varies, no uniform plan is evident in all these gates. Some gates are huge, complex and spacious

while some simple and small in size. The height of this gate including parapet is 25 feet and the span of the porch is 8 feet. The gateway has two pointed arches. The plan of the gate is simple. The porch has chambers on either side. The ceiling porch chamber are vaulted dome style. The gate is also known as 'Suttari Darwaza' or 'Carpenters gate'.[22]

7. **Katgar Darwaza :** This gate's height is 35 feet and is in close proximity to the Shakar Baoli. The gate appears majestic and is flanked by two bastions. It has a domed ceiling. The porch is 15 feet with two chambers on either side and the ceiling is covered with stucco work of diamond cut design. A lion is carved as if roaring. The gateway is flanked by semi-circular bastions with parapet and machicolation.

8. **Delhi Darwaza :** The gate is also popular as Rauza Darwaza and is 12 kms. away from Daulatabad. Since it faces towards the north it is known as Delhi gate. The total height is 30 feet. The porch is dome shaped and has two chambers on either side which are elevated 5 feet above ground level. The ceiling is again vaulted—unlike other gates this gate is devoid of a towers or bastion.

9. **Khas Darwaza :** The height continues to remain at 30 feet including the parapet wall. The gateway has two soaring arches the space between these two arches measures 20 feet. The gateway also has two flanking semi-circular towers on either side. The bastions are also provided with machicolations. This gate is the last gate of Ambar Kot and ends at about 400 feet away from this gate. The wall joins from the back side with the outer fortified walls moat and disappears thereafter.

Inner Defence Fortifications of the fort :

The medieval period was an insecure period especially as consistent fear of invasions not limited to the neighbouring kingdoms but also from distant lands must have compelled the rulers to construct and create the most secure and permanent defences as protections against surprise attacks. Insecurity reigned and strong defences were a compulsion.

Daulatabad fort was strategic and hence required enclosures of cyclopean dimensions. The inner defensive wall was probably rebuilt during the Tughluq period encircling the expanded town. Defences were made complicated by constructing four distint lines of walls with innumerable bastions and towers, which would have to be scrambled before the main entrance could be reached. The average height of these walls is between 20 to 30 feet and is 6 to 8 feet in thickness. These walls however ended at the moat and joined the moat on the north and south side of the fortress.

Mahakot fortification first line defence :

The outer wall of fort (Mahakot) is two or three quarter miles in circumference.

This wall enclosed the entire city and was built around the base of the fortress and which can still be traced by cluster of ruins embosomed in perfect wilderness of custard apple trees and jungles. Some portion of the wall is in a fair state of preservation but some portion especially facing the present road is in a bad condition due to continuous vibrations of vehicles. It has loop holed battlements with circular towers at the angles on which occasionally pieces of antiquated ordance is visible.[23] The moat is very wide and measures 80 feet in width and 20 feet in height. The present condition of the moat appears to be in a bad shape. The water has receded, and the remaining water is choked with dirt and filth and the exterior portion is covered with dense forest area and forms a congenial lurking place for wild birds and animals.

The modern exterior or the main entrance to the fort is through a tall walled lane with a descent into the open space. The first wall or the outer wall is faced by broad deep moat without water. Earlier it must have contained water. This outer wall is in a bad condition as on either side of the main entrance an open parking space has been allotted for the tourist vehicles. The continuous movement of the vehicles has affected the outer wall the most as it has not only reduced the height of the wall and caused erosion but the wall still continues to loosen the stone bindings.

The entrance in the open space is followed immediately by another fortification wall. The outer wall still contains heavy iron chains which when the place was besieged could be stretched from wall to wall and so prevent elephants from affecting an entrance. Two strong teakwood beams were also used for the same purpose[24] (Plate II).

The interior fortification wall is followed by a succession of walls and gateways and courtyards. The walls are massive thick and convolutes on the outer side faces and is defended by large bastions both without and within the court. The fortification wall of Mahakot consists of 54 small and big bastions.[25] From the south east base of the fort to Mahakot Darwaza there are 32 bastions and from Mahakot gate to the end of this wall there are 22 bastions. The bastions are either single or double storeyed and few are three storey bastions. The length of the wall from southeast to mahakot is 4800 feet and from Mahakot to end of the wall it measures 4000 feet.[26] The north side huge bastions is known as Badshah pasand and where the wall ends the bastion is known as Gauhar pasand.[27]

The wall towards its right has a huge gateway which consists of a teak doorway with iron studded and spiked nails to prevent elephants from battering them. They are flanked with round towers about 40 feet high and are closed by strong bars of teak. On the right of the entrance gateway, is a huge bastion, the upper part is lock hold for musket fire. The face of the gateway above the doors has been pierced with three large openings for artillery and the parapet over it has been rebuilt with level coping[28] (Plate III). On the upper side above tiny groups of lions and elephants are carved in stone. The lions signifying safety and security of the building according to Shia belief.[29]

The gates open into a huge capacious courtyard by rows of guard-rooms for stronger protection. Now these guardrooms lack doorways and display canons and guns several with animal figures carved on them belonging to different periods and dynasties. The earliest probably belong to Aurangzeb's period as he spent a major part of his life in the Deccan. Some canons belong to Nizam of Hyderabad and some to Malik Ambar or the Nizamshahi state of Ahmednagar. The measurement of the courtyard is 80 x 80 feet and this courtyard again is closed further by another gate with a fortification wall. This gate has a lofty height and its architectural details reveal Hindu influence. The towers of the gateway are very high and is defended by strong towers and an embattled parapet (Plate IV). There are two leaved doors armed with iron spikes. The porch has two small chambers on either side. The chambers are supported by small pillars of Hindu construction and probably belonging to a temple. There are panels of trees and flowers. A flight of steps leading to the top is also evident (Plate V).

Parkota :

This is an additional measure of defence and is applied to such barbican as are built below a bastion (kot) guarding a passage. This defence management at every step creates obstacles for the enemy and deters his entrance making it impossible for him to enter. The movement of the enemy is stopped and repressed at every point. The entrance is fettered and exposed to the fire of the garrison. The interior of the parkota has been converted into lawns and the fortification walls are bordered by jain sculptures remains of the temple (Plate VI). The area between the first and the second line of the references on north side is also known as Parkota. Its length and width varies from place to place.[30] On the eastern portion of the second line is a gate known as 'Parkote Ka Darwaza' leading to the Parkota.[31]. This is arched gate also spiked and studded a device against elephant attacks. On the north side of the Parkota there are guard rooms and its western wall contains a two storeyed bastion massive and majestic a place from where fire could be opened. Main purpose is additional defence management. The height of the bastion offers a good view of the surroundings and any movement of the enemy from a safe distance.

Barbican :

Generally refer to tower gateways. This was a muslim innovation of bonding columns inside masonry in order to strengthen.[32] This was a strong protective defence. These barbicans again besieged them from three gates successively each facing a different angle or zigzag position, and each exposed to a bewildering array of vantage points one higher than the other.[33]

The second line of Fortification wall and tower :

After the second gate the road to the third turns abruptly to the right and traverses a distance of about 80 feet before the next line of defence is reached. To the left of the gate is a large semi-circular conical tower styled as Naggar Khana[34]

(Plate VII). The tower is in a bad shape and broken but in its original condition must have commanded a view of a large portion outside the fort. To reach the following gate in the hornwork one must pass diagonally through the court exposed to attack from all sides, this gateway, closed only by a single two leaved door is much narrower than those already passed being 9 feet 5 inches jamb as compared with 11 feet for the gateway at the entry to the second court and 12 feet 2 inches for the doors at either end of the outer passage[35]. Beyond the gateway is another smaller entrance which leads within the limits of the lower portion of the fort. A roughly paved ascent some distance in length running between two stone walls five feet high leads to the next portal inner defences. The second fortified wall begins from the south east and has 21 bastions, on the northeast 9 bastions, to the north west the wall measures 4000 feet. From the third gateway a narrow road leads to the fourth gate.

Kala-kot :

This is also known as Kalagate. One has to walk through a narrow road for a distance with guard rooms on the right, (Plate VIII) further the majestic Chand Minar can be sighted. In the beginning on the left is the huge Hathi tank (Plate IX). The walk is through a cobbled roadway.

The Kalagate also has spiked defence and strong iron bars :

The gate is so strong and powerful that it appears to be indomitable and a Herculean task to break it open. The entrance on either side has raised platforms and guard rooms. The roof of the gateway is partially ruined and is flanked by two towers upon which are bastions for guns. A broad flight of steps leads to this gateway. There are recesses directly in front of the enemy and his further progress is obstructed by a door on the right.

As the flight of steps move in a curvature one is confronted with several gateways. There are at least five to six gateways within a short distance with recesses for storage of ammunition (Plate X). The sixth gate has curiously shaped oblong porch with semi-circular roof. Opening to a passage through the wall the flight of steps contain large recesses. An ascent of about 60 to 62 steps leads to the seventh gateway (Plate XI). This gate appears older than the rest of the gates and its teakwood has blackened with age. To the left of the gateway and adjoining to the entrances one can witness palaces of the earliest period of the Yadavas, presently in a ruined condition.

As one walks within this wall a huge courtyard is witnessed with the Chini Mahal facing the gateway. To move towards the next gateway one has to turn to the left. This gateway is also defended by bastion and towers and mounted with guns. On the top towards the right on a slightly raised bastion one can witness the Mendha top so called due to its raised carving of a ram on the canon. This open courtyard overlooks the entire vista beyond the fort. From this point one can also appreciate the skillful manner in which the hill has been scarped, and around upto a height of

150 feet. [36] Even in the 21st Century despite torrential rains of centuries, extreme summers the vertical face of the rock is practically as smooth and unassailable as when first scarped. The escarpment is so smooth that neither a snake nor ant could scale it [37] The hill which is scarped is 3500 feet in its circumference. The moat is 50 feet deep and was always filled with water. The height of the water has however receded considerably. The water today reeks of slime and slush and is completely contaminated. Very recently the ASI has cleaned the right side moat completely.

Some of the medieval monuments also were defended by deep ditches and if situated on the bank of rivers they were defended by the river on that side and ditches on the others[38] as the ditches being crossed at the gates by drawbridges as at Agra. When on a hill the outer walls were often so defended as at Ginge and Golkonda.[39] At Bidar there are three ditches around the city defending the fort. At Daulatabad at the main entrance to the fort, there is a ditch and the other sides are defended by high walls.

Moat and Bridges :

Ibn-e-Battuta who visited Daulatabad during the period of Muhammad Tughlaq writes that 'the rock has been excavated and a castle was built on its summit. It is reached by a ladder which was taken up at night.[40] Later in 1874 A.D. the bridge was reconstructed during the Nizam Asif Jahi dynasty of Hyderabad on the orders of Sir Salar Jung.[41] This bridge is the only entrance to the main fort. After the moat is crossed than a flight of steps nearly twenty in number lead down to the bridge and passes from bottom step. The bridge is 6 feet 8 ins. The gate of this bridge is very narrow measuring only 2 feet 5 inches in width and 5 feet 6 inches in length so that it is difficult for an enemy army to enter (Plate XII).

The height of the water in the moat was under control and any adjustment in this height could be restricted to the space between the dams (Plate XIII). During any siege of the enemy was conducted if the enemy did succeed in crossing all barriers the bridge could be flooded and the water in that portion raised to such a height as to render the bridge impossible or atleast to make the crossing under attack impracticable.[42] The present bridge which is witnessed today was re-constructed in 1952.[43] After the bridge is the gallery which passes around three sides of a tall bastion erected by Aurangzeb.[44]

Dark passage :

The road leading to the entrance of the gallery through which admission is obtained to the upper portion of the fort skirts the base of the bastion until it reaches a small teak doorway which gives entrance to the first gallery. This door is around 6 feet in height and 4 feet in breadth. The length of the first gallery is 60 feet long and leads to a small roughly constructed courtyard (Plate XIV). In one corner of this courtyard is a broken flight of steps leading to a small plateau containing the ruins of a medieval palace.[45]

The roughly constructed courtyard, already mentioned, the second gallery, is entered by means of a flight of ten steps. However to penetrate this portion of the ramp one has to be provided with lanterns or torches. Generally the watchmen provide the tourists with this facility. The height of this passage is around eight to ten feet and is about 30 feet in breadth and 60 feet in length terminating in a small vestibule ornamented with pillars[46] Beyond this is a long tunnel containing numerous chambers cut out of the solid rock which were used in the olden times as guard rooms and store houses.[47] At the end of this passage is another flight of steps leading upto a small place over looking the moat, a further ascent of 43 steps leads to a small square opening about 30 x 20 feet which forms the termination of the tunnel. At the head of the tunnel is an iron shutter measuring 5 feet 6 "x 6 feet 6" (now raised from its original position), This runs horizontally on small wheels, covering or uncovering the opening like a trap door. The part of the original cover is lost.[48]

Provision of a barrier of smoke was one of the most ingenious and effective defence technique of this tunnel. The shutter is ½ "in thickness, which was used for barring the ingress of the besiegers. The plate was laid across the opening and huge fire kindled upon it,[49] so that it became red hot and rendered all approach to it from the inner gallery impossible. At a point about half way through where the tunnel passed, here the vertical face of the rock hole measuring 5 feet in diameter was cut through for the fire in an iron brazier which was installed in a small chamber opening into the tunnel. When the fire was kindled the current of air from the hole would waft the smoke up the tunnel and render its passage impossible. A somewhat similar device described by Polyblius was adopted during the siege of Ambracia in 190 B.C. to expel the Romans from the mine they had driven towards the city.[50] The only entrance (to the upper fort or inner citadel) is through a devious tunnel which in times of siege was rendered impassable by an ingenious contrivance was the description given by Cambridge History of India. It further writes 'At the bend in the tunnel which came near to the outer edge of the rock was a small chamber provided with a flue pierced through the thickness of the wall and fitted in addition with a staging of iron plates, of these plates a charcoal fire was fit, which was fanned by the wind flowing through the flue, would quickly fill the tunnel with its fumes and make any ingress impossible.

Bastions :

The technique to strengthening the fortification walls by means of the use of half round projecting towers had its origin in Baghdad in around the 12th Century A.D. This technique was used to a great extent and rampantly at Daulatabad fort in around 1327 A.D. When the colossal construction of Daulatabad was raised. All around the fort such massive bastions are evident. While constructing the fortified wall around Daulatabad town and fort most advanced techniques were developed in the construction of bastions[51] (Plate XV).

Fig. III. Horn Work.

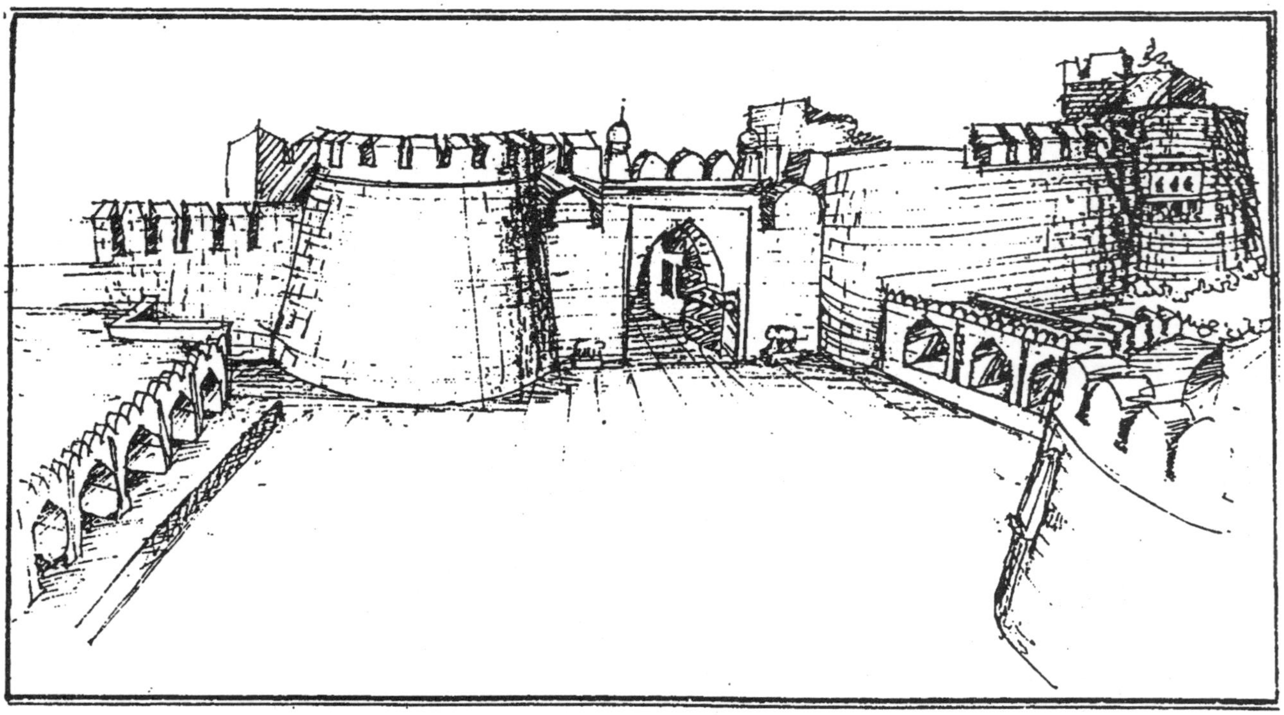

Fig. IV. Horn Work with open courtyard.

Fig. V. Dry Moat.

Fig. VI. Wet Moat.

Fig. VIIa. Bastion's and fortification wall.

Fig. VIIb. Bastion's and fortification wall.

The bastions were raised in the structure of the curtain wall and other encircling wall of Daulatabad fort with an average interval varying 60 ft. to 160 ft. Heavy cannons or piece of ordance were mounted on the top of the double storeyed bastions constructed to protect the main gates of the fort, of town as well as the fort. On top of these bastions heavy, formidable looking guns were placed in order to carry specially large pieces of ordnance. Among these bastions the famous Kala Pahad, Mendha top and Jahaz top are carried. The names of the bastions are salami Burj Fateh Aurang and 'Mang Burj'. Indeed the entire walls and bastions provide the best example of military engineering skills.

The Mendha top is mounted on a large, high bastion (Plate XVI). In the center of the paved platform small circular holes to hold the pivot upon which the carriage revolved and at a distance from it decided in each case by the length of the gun are opposite segments of a channeled ring. Cut into the pavement in which the wheels of the carriage travelled as the gun swung around. It appears that in the original construction of these bastions no cover was provided for the gunner. The other bastion nearing the moat is octagonal in shape with a height of 40 feet erected by Aurangzeb for the reception of the huge piece of ordnance. Presently it only contains a small iron gun without any inscription and is now half rusted.

A bastion called 'Fateh-Aurang-burj is built within the fortification wall of 'Kala-kot' on the south side of the wall. Near the bastion a room for storage for gun powder was also made. Each and every bastion lying within the fortification wall of Daulatabad fort (Mahakot and Kalakot), are having rooms for storage of gun powder.[52] This also suggests that every bastion of Mahakot and Kalakot were mounted with heavy or small pieces of ordnance.

The fort is also strengthened by double storeyed bastions, massively built. A bastion near the main entrance is double storeyed and is supported by huge pillars. A passage has been built around the pillar to allow free movement for gunmen to fire from loopholes.

The most famous bastions of the outer most walls are however known as Salami Burj and Mang Burj which are mounted by heavy pieces or ordnance.[53] These bastions are double as well as three storeyed. Another interesting feature of these bastions are the wells that are dug or provided within the mason tank within the bastion.

The best feature of these bastions are the balconies on the upper storey. Two storeyed tower with covered chamber in the second storey and an open platform protected by a parapet on the roof resulted in a greater and more effective concentration of fire power.[54]

Darwazas (Gates) :

The medieval gates huge and strong provided powerful protective defences to the fort. A lot of attention was paid to the major gates and drew the special attention of military strategy. Most of the simple gates were made up of strong teakwood and were closed and barred from inside.[55]

The main gate of Daulatabad is based on the technology of construction of pointed horse shoe arches started in Baghdad by Qasr-Al-Haur.[56] The massive construction of Daulatabad also depict medieval innovations which developed in India. The gateways sometimes are so powerful that all the strength and military strategy fails to break open the gates. They sometimes take the form of two powerful walls that extend out beyond the gate with towers at the end of a sinuous road being defended by box machicolations, jutting out from the parapets as at Golconda and Bidar. At Daulatabad the outer gate has two doors, one at entrance and one at exit into the first court with vaulted passage in between (Plate XVII).

In most cases the passage through the gateway occasionally has a door at either end but at Daulatabad, normally there is only one door and that too at the entrance as at Lahore gate of Delhi and upper gates at Agra.[57] The doorways are very large generally from 12 to 16 feet wide and upto 25 feet high to the point of the arch or underside of the lintel. The width is not much greater than that of Roman and medieval town gates of west.[58] The iron spikes attached to a door which was generally of teak. Iron was arranged in horizontal tiers across its outer face. These doors as described earlier could not be pushed open by the most powerful enemy even with elephants.

The exterior most wall of the fort measures nearly three miles and is pierced with nine gates excluding small postern wickets. The small openings can be seen in all periods of Greek military architecture and served a variety of purposes. The great majority of postern were placed in the wall of the outer city and gave direct access to the open courtside.[59] These nine gates and several window postern at the fortified walls have been described by Raunaq Ali.[60] The names of these gates and their descriptions have already been provided in the earlier pages.

The names of most of these gates especially the more significant ones are not only found in the Deccan but also Delhi. Names like Delhi gate is found in Delhi while names like Mecca are found in Aurangabad, Bijapur and Golconda. The Paithan gate is associated with Aurangabad's south gate while Rauza gate to Khultabad. Khadki again is associated with Aurangabad and the Fatehabad gate faces Fatehabad which is 5 km. away from Daulatabad and hence bears this name. Sometimes gates were also named after a particular community like 'Suttari Darwaza'(Carpenter's door)[61] of Daulatabad. At times the gates were named after important and politically influential people like Islam Darwaza, Zafar and *Mahmad* Darwaza.

Battlements :

Battlements are parapets having a series of indentations or embrasures, between which are raised portions known as merlons; archers or musketeers could shoot through the embrasures between the protecting merlons.[62] These are very good techniques of defence seen at Daulatabad fort (Plate XVIII). In the fort different kinds of techniques were used keeping in view the need of defence. Arrow olits for downward shooting are found in the European Citadels.[63] Loopholes were employed here and in Daulatabad the loopholes are generally two tier loopholes, the upper tier piercing the merlons and the lower dipped rapidly down from the wall walk to appear on the enterface far below the parapet, the holes, through sometimes opening but laterally at the back, were often parallel straight through.[64] At Purana Kila (Delhi) there are three tier loopholes one from an internal wall arcade and two from the battlements. At Tughlaqabad there are four tiers, one from external gallery, one from the internal gallery and remaining two from the battlements.[65] The parapet walls are made up of huge blocks of stone bound by lime plaster upto the wall walk level on the exterior and interior faces. But for the battlement more intricate work was required hence stone blocks could not serve the purpose hence bricks were used which were constructed to allow small square holes above which are parallel through the parapet and meant for defence by musket fire, barrage etc. The battlement also made special provisions for sighting and below large holes opened out for canons.[66] The outer face of the loopholes is covered with small stone hood for the defence of a man firing through it as at Bijapur, Fatehpur Sikri and Agra.[67] The loopholes constructed at the various bastions of Daulatabad fort and at the merlons are very large in size.

The fort walls not only stretched lengthwise but for defence purposes its breadth too was broad so as to serve for a comfortable walk. In all the forts, like Agra, Delhi's Red fort, Chittore or Amber fort wall walk are its common feature. On the curtain wall of Daulatabad, there is a wide crenellated wall walk which runs all round the curtain from bastion to bastion, the wall walks were 4 feet to 5 feet wide and served as a sound defence strategy.

Cave temples of Devagiri fort :

Within the bosom of the medieval fort the ancient imprints of the past in the form of cave temples are witnessed even today though in extremely ruined conditions. They are located on the eastern ridge of the fort, around 400 steps from the Kala kot Darwaza. The entry to the caves is blocked today by a thick growth of wild bushes and trees. As one surpasses this growth ruined debris are visible which reveals the basement of a huge palatial building now in a bad state of preservation. This basement provides the approach road to the rock cut entrances to a number of caves. The total number of caves are five again in a bad condition.

DESCRIPTION OF CAVES

Cave No. I :

This cave is situated on the extreme left and is square in plan with a pillared hall. The back wall has cells. The pillars are huge with square base, squarish shaft crowned with an octagonal compartment topped with a square capital which is carved with floral motifs in semi-circles.[68] The pilasters are massive but simple in

CAVE - I

CAVE - II

Fig. VIII. Cave line drawing.

design. In front of the cave is a cistern also cut in rock. In immediate proximity stands an unfinished cave blocked by stone wall. The cave has no sculptural remains. The measurement of the cave are 9.25 mtrs. x 9.70 mtrs. x 2.30 mtrs.

Cave No. II :

This cave is also pillared with three cells in its back wall. On its extreme right of the backwall is a disfigured passage leading to the moat. The pillars and pilasters of the cave also have square base and shaft, topped by constricted cushion and Ghatapallava motif. The capital of the pillar is crowned with nagabandh. On its front side is a small cistern. There are no sculptural remains in this cave. It measures 13.80 mtrs. x 10.20 mtrs. x 2.85 mtrs.[69]

Cave No. III :

This cave appears to be well-planned similar to other caves of Ellora with a pillared hall, an ante-chamber and interior shrine (garbhagraha) in its back wall. The front of the cave is flanked by a row of kakasana on either side. The pillars are again square in shape, with a shaft and surmounted by an octagonal part decorated with floral motif. This portion is again topped by kirtimukhas, yogis and birds. There are several niches on the walls of the ante-chamber. The doorway leading the inner sanctury is decorated by pilasters, jambs crowned by the ghatapallava capital, bordered with kani-mouldings. Within the shrine the pedestal still is visible devoid of sculptures.

Cave No. IV and V :

These are very small caves close to cave No. III. These caves have similarities with Kailash which is at close quarters from the fort.

Beneath these caves are huge dungeons referred to by Ibn-e-Batuta and probably used for storing purposes. The caves probably contained Jain sculptures and belonged to the Jain origin. The contention that they belonged to the Tantrik cult would be far fetched assumption. These caves are pre-Yadava (Plate XIX).

Temples and Mosques

Jain temple dedicated to Goddess Kali :

This temple is situated towards the left of the minar and was converted as a mosque and used till before the independence period; since it was under the Nizam of Hyderabad. It was an active mosque and was also used for festive occasions like Ramzan and Bakri-Id.

Jama Masjid cum Bharat Mata Mandir :

One of the biggest temples which during Qutubuddin Khilji's period was converted into a mosque. It was used as a Jama Masjid and is capacious and impressive. The number of pillars which probably once supported a corridoral ceiling has vanished. The pillars within the main temple still show obvious imprints of Jain temples. There is evidence that the first king of Bahamani dynasty Alauddin Bahman Shah was crowned by his preceptor Shaikh Sirajuddin Junaid on Friday 24/4/748/3/8/1347 A.D. in the mosque of Qutubuddin Mubarak Shah Khilji at Daulatabad.[73] But after the Muslim power vanished the mosque was again converted into a temple and now is known as the Bharat Mata temple (Plate XIX).

The Bharat Mata temple lies towards the left as one walks the narrow lane between high walls towards the Kala Kot and occupies a huge quadrangle space and is aesthetically graceful. The entrance gateway (Plate XX) is in the center of quadrangle and the other three sides also contained similar gates and cells as their remains are still evident. The main entrance is towards the west and is topped by dome adding to its beauty. The porch contains two pointed arches on either side. But the north and south gates are original. Across the west side of the huge courtyard, is the mosque made up of pillars of Jain architecture. This portion is expansive and measures 250 feet from north-south and 40 feet east west. Nearly one hundred and fifty eight pillars support the roof and all are carved. The brackets and cornice is ornamented with lotus flowers and other geometrical designs and figures of the Hindu deities can be identified (Plate XXI). The central back wall now contains the Bharat Mata Goddess.

Centrally a huge dome is witnessed an obvious architectural feature of the Muslim period which was added later to an otherwise flat roof (Plate XXII). Very similar dome is evident at Delhi in Alai Darwaza in Qutb Complex and other structures of Khilji period.[74] The interior pillars within the roofed structure of the main temple are in a good state of preservation but the rest of the pillars spread on all three sides of the open courtyard are in a bad condition. The pillars which are spread and constructed all round the temple on the three sides and several of the pillars in the main temple display varied workmanship and are not similar in design or structure and hence it is obvious that they are parts of different temples.

The mihrab has been exquisitely finished and effectively set into the plan of the wall lintel, and have carved design. As a whole the nave of the temple has been most elegantly finished, the pillars perfectly symmetrical giving an impression of a harmonious and well adorned temple. The squinches have receding planes with fringe of flower buds, fringes have also been used on mihrabs and along all arches and alcoves. During the Mughal period too mihrab was used as a decorative feature in the mosque with additional embellishment like mosaics, glazed tiles, colored glass, colored stones etc. The other places where singular mihrabs are evident are at the Lal Darwaza mosque of Jaunpur 1440 A.D., Atala mosque again from Jaunpur 1375 AD., and Qila-e-Kuhna masjid of Delhi.[75]

Another notable feature of the mosque was the cylindrical pillars which serve more or less as buttresses, standing at the end of walls which have a series of arches each with an outward thrust [76] To construct an arcade, a series of arches on western side facing the courtyard as a facade to sanctuary for which indigenous artisans were employed. The ogee of these arches too came through Hindu buildings and has no relation with pointed arches of Muslim monuments. Thus the construction of the huge temple, later mosque and again temple was a complete re-arrangement, alteration, consisting of lending a new order to the pillars according to their own requirement. It was a beautiful blending of Indo-Islamic architecture of well proportioned artistic Hindu pillars changed to a new order, the plan slightly changed with additional Muslim architectural features of arches, arcades and domes which they built with brick mostly with temporary centering.

The author of the temple has not yet been identified except for various local traditional versions that it was built by Raja Ramchandra the Yadava prince of Devagiri but this has no proof[77] Another tradition gives the credit to the daughter of Raja Bhoj within six years.[78] And the latest record gives the credit to Prithvidhara or Pethadakumar a rich merchant and minister of the Paramara of Malwa[79] as referred earlier.

A temple and Mosque near Kala Kot :

Around fifty feet from Kala Kot gate towards the north is another pillared temple converted into a mosque. It is smaller in size and has a central clerestry rising from the roof. Hindu architectural features of carved pillars with lintels, brackets and flat ceiling is still evident. Presently the mosque measures 18.80 meters south north and 10.50 metres east-west. The roof is supported by a series of square carved pillars of which there are 60 altogether. The height is around 10 feet. The mosque according to one version is the creation of Muhammad Bin Tughlaq[80] for which no authentic authority has been quoted. However the specific architectural features bear similarities to some of the special features of Muhammad Bin Tughlaq' s architecture and hence the assumption that probably he was the patron of this mosque.

The pattern of construction of the mosque is similar to the Bharat Mata temple but it is much smaller in size and contains neither a dome nor a big courtyard. Its eastern side is completely open and has small courtyard. It has an entrance on the southern wall. The mihrab too has beautiful designs and is set into plain wall lintels. The Islamic talisman for strength and victory has been effectively depicted on mihrab and arched niches. The decorative motifs are typically like lotus and geometrical patterns.[81]

Bhavani Temple :

A short distance to the right of the third gateway are the remains of what once is an elegantly constructed Hindu temple dedicated to Goddess Bhavani.[82] The central part of the temple is occupied by a flight of steps leading to the roof of the temple which is flat. This was constructed on pillars and lintels like that of Bharat Mata and a temple dedicated to Kali. Ten rows of carved pillars support the temple.

Dargah Shah Aziz Quattal :

Aziz Quattal was a Sufi Saint and the younger brother of Hazrat Zar Zari Baksh and Hazrat Burhannuddin Garib. A narrow pathway from the Chand Minar and the Diwan-i-Am leads to the Dargah. The dargah is in a bad condition. Towards its west is a mosque. Near the dargah is a big and it is from here that Aab-Pash Dara nehar turns towards the fort area, which joining to the Hathi tank.

Step Wells :

The step wells of Daulatabad offer a rich variety of the Hindu water system. The Hindus since ancient times were fond of raising cities on the banks of river tanks or wells as water held religious as well as purifying connotations.[83] There are certain surviving evidences of step wells and also references of such wells at the religious sites of Ujjani, Dwarka, Ayodhya and Benaras. The Hindu vastu had a definite place for such wells and artificial lakes like Kurukshetra must have required ingenuous skills of the architects to confine and contain the water which formed one of the greatest attractions of the images of gods and goddess in the Devalaya with Vapi (baoli), Kupa (well), Tadga (tank) as the virtuous act.[84] The construction of such wells was considered an act of merit.[85]

According to the Hindu architectural cannons there are four types of wells mentioned as Nanda, Bhadra, Jaya and Vijay.[86]

The most notable survival of the ancient water systems are evident in Gujarat and monumental wells of Gujarat belonging to the Hindu period are mentioned here.

The step wells (baoli or baodi) consists of two parts, a circular or an octagonal well (called its shaft), from which water is drawn up either by pur (leather bags) operated manually or by bullock or by a series of galleries or floors, connected by flights of steps descending successively from the uppermost platform to a lower landing on which stands pillars supporting a roof of equal breadth above. A second flight of steps continues stage by stage till the water level is reached.[87] Round the shaft of this reservoir are platforms in each storey protected by parapet surrounding the octagonal shaft their backs forming seats for loungers.[88] The Ahmad Shahi dynasty of Gujarat witnessed a perfection in step wells. Examples of these wells are at Bai Harirs Wav at Asarwa built in 1499 A.D. and the Wav of Adalaj about 12 miles from Ahmedabad, built in 1502 A.D.[89] Step wells were seen in most parts of India and are pre-Mughal.[90]

Step wells of Daulatabad :

Step wells of Daulatabad have immense artistic and architectural value. Four step wells are prominently seen presently.

Dargah Momin Arif Stepwell :

Near the Dargah situated near the tomb of Hazrat Shah Sayed Momin Arif measures 80 feet in length and breadth and 40 feet deep. Its sides are paved with square blocks of stone and have narrow flights of steps leading to the water level. [91] The water of these baolis comes from natural springs. In this well there are a number of flight of steps, galleries and platforms.

Saraswati Step well :

This is a well dedicated to Goddess Saraswati and is situated in the interiors of the fortified walls. It is about 50 feet away from the third gateway, near s Bahlul's tomb. It is a huge well measuring 100 feet in length and breadth and 40 feet deep. Sacred to Goddess Saraswati has paved sides with square blocks of stone and has narrow flight of steps leading to the water level. The water source are pipes connected with a reservior to the north east. Presently the pipeline probably at some places is disconnected hence there is hardly any water supply. An inscription on the outer wall of the tomb of Qattal Shahid refers to Ghiyasuddin Tughlaq Shah the Sultan of Gazi in the month of shawwal 722 A.D./1322 by one Thakkar Nanak of Jagbir.[92] The inscription also records that the well was constructed for providing water to the common people and mainly for public use[93] (Plate XXIII).

Step well near Kabdi Gate :

Outside the wall of Ambar Kot is one of the most beautiful among all the step wells of the fort and is in a very good state of preservation. Its sides are paved with square and quadrangular block of black stone and have narrow flight of steps leading to the water level. The steps are only to east and west sides of the well. The step well contains, platforms, flight of steps and galleries. There are two sources of water one from natural springs and the other from rivulet that runs close to the well.

HAMMAM BATHS

A historical sketch :

Generally the very mention of the word Hammam brings to mind the Turkish baths, the elaborate fashion in treating the body to a relaxed environment. In Arabic the term Hammam means heat or temperature as around hammam there is heat at one place and cold at another and hammam is such a place where heat or cold is created to effect the body of individuals.[94]

The earliest traces of such elaborately treated baths can go back to as early as

Fig. IX. Hammam.

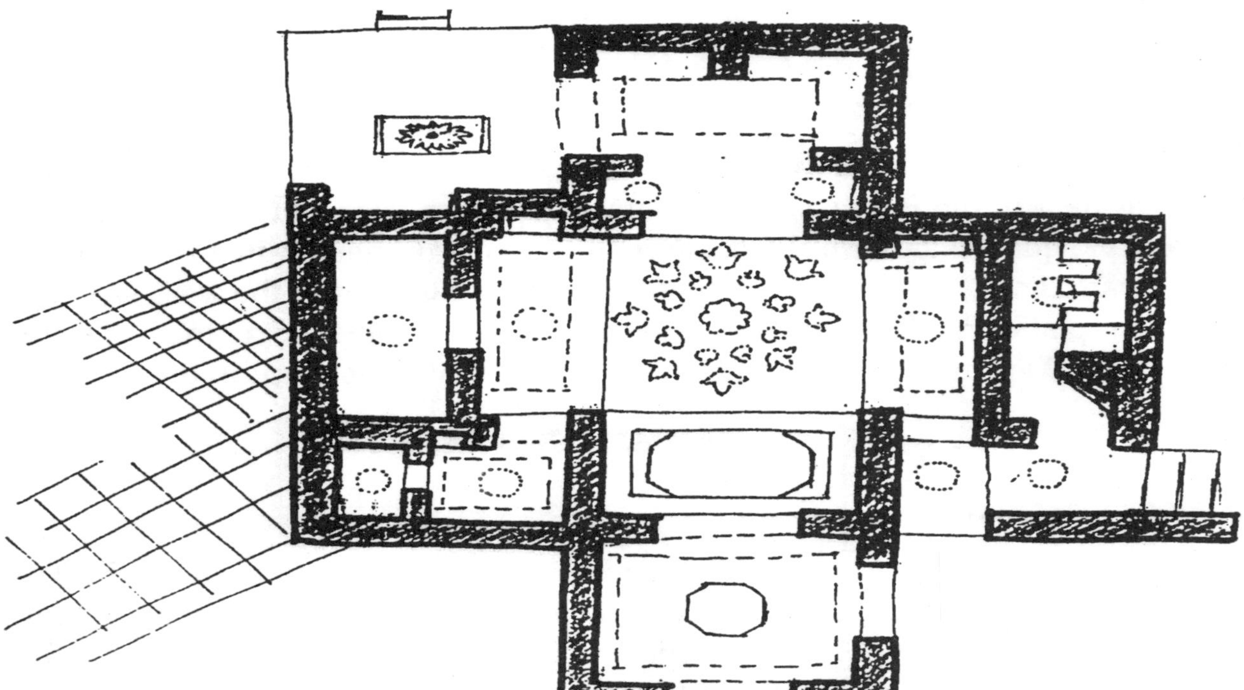

Fig. X. Plan of Princess Hammam.

2500 B.C. at Mohenjodaro and Harappa wherein the Indus Valley civilization seem to have given a lot of significance to baths as the most elaborate public baths of huge dimensions were constructed, with rooms all around. Built mostly in burnt bricks the binding was lime and there was very little scope for water to escape or leak from the strong brick constructions. After this period such huge baths were not evident and instead the practice of constructing a large number of 'kunds' within the precincts of the temple became common as a worshipper had a bath before entering the premises of the temple.[95]

Ancient civilizations like Assyrian and Aegean periods elaborate examples of baths have been found. At Knossus and Phaistos (1700-1400 B.C.) and at Tiryana (1200 B.C.).[96] The Roman baths are some of the most advanced and high wrought specimen of baths and naturally due to their delight in physical luxuries developed the bath on sumptuous scales.[97] Seperate baths for private and public purposes were constructed and were known as 'therma", which is synonymous to bath.[98] The bath at villa of Diomed was for private use while the stalean bath and Forum bath were open for public use.[99] The important feature of the Roman bath was the hypocaust which is a system of heating a room to high degree developed for both houses and public bath.[100]

The Roman technique soon showed its impact on other neighbouring countries as far as Persia, Iraq and Asia minor. As Islam spread a new culture was introduced and the awakened and enlightened muslims adopted and developed the ancient Greeko-Roman technique of bath.[101] The descendents of Mauviya built steam baths at Damascus in the Roman-Syrian fashion.[102]

The special feature of these baths is a barrel vaulted room maintaining a tepid temperature. These baths were ornamented to give a luxurious effect. Full breasted birds with combed heads and dainty in a Plumage turbanned busts arrayed in a series as in a pertrail gallery, atlas figures bearing up the dome of the bath entrance.[103] The baths in each age were developed and improvised and facilities for hot bath, cold bath, vapour bath, acid bath, Turkishbath, oil bath, sandbath, sun bath, foot bath, mud bath, feet bath, medicated bath[104] and a horde of other baths were developed and popularized.

Hammam of Daulatabad :

Since the rulers in Delhi during the 13th Century arrived from the muslim countries they brought with them their ideas, habits, culture and practices. They introduced the hammams in their palaces. Similarly when Muhammad Tughlaq transferred his capital from Delhi to Daulatabad in the year 1326-27 he also transferred all his population to the fort which necessitated the creation of an entire city with all amenities and facilities required by the royalty and as was created in Delhi. Just as the entire way from Delhi to Daulatabad was studded with sarais for resting purposes so also the fort premises were provided with all the similar amenities as in the royal palaces of Delhi. Hammams were an important feature of these palaces. According to Iqtidar Hussain and Qazi Ahmad, the entire city was divided into colonies according to their occupation and craft and each of these colonies was provided with all the amenities like a separate bazar, mosque, minarets, ovens for flour mills and public baths to avoid dependence.[105] Such public baths have also found mention in Ibn-e-Battuta's records who visited Daulatabad during the period of Muhammad Tughlaq, in fact, accompanied Muhammad Tughlaq. This leads to the conclusion that there were several public baths at Daulatabad.

The appearance of these hammams from a distance was not very attractive and had no attractive architectural features. But internally they were extremely comfortable, attractive and soothing. Seperate baths were provided for ladies and gents and the interior designing of these baths was according to the tastes of each sex, mostly floral designs decorated the baths with stucco and painted with various colours and made as pleasant to look as possible.[106]

Daulatabad fort has been endowed with a number of baths some still in a very good state of preservation while some are in complete ruins. Some of the baths are wonderful examples of completely developed baths. Riyaz Ansari for the first time has dealt at length with the description of the hammams in his thesis, though by this time a large number of hammams were already in either ruins or in a bad condition. However some hammams are still in a good state of preservation and can be studied at length. Though these hamams too if not conserved will soon fall in complete ruins. Archaeological Survey of India generally believes in merely white washing the old monuments instead of giving it a proper well deserved face lift. As far as Daulatabad's baths are concerned they have neither been repaired nor white washed but allowed to fall in ruins. The ASI has given it complete liberty to deteriorate.

The baths now existing at Daulatabad which are in a good state of preservation permit us to study its architectural details. The hammams built in the Greeko-Roman style display their paraphernalia furnaces, water channels and earthen pipes. These baths are also evidence of the fact that such developed, luxurious baths existed event before the advent of the Mughals in the Deccan and even it's technology was advanced. According to Riaz Ansari[107] the baths in such a developed state arrived in the Deccan for the first time in India and it was only much later that North India also followed suit. Of course this is an assumption which has to be studied with proper comparisons both in the north and eastern parts of India where too the muslims had made inroads much before the arrival of the Mughals.

Baths in India again according to Riaz Ansari[108] are of two types—the Deccan style and the Mughal style. The Mughal style of Hammam architecture can be considered as some of the most excellent, sophisticated, model examples of luxurious, sensual baths. The best example of Deccan Daulatabad can be seen at Bidar.[109] The best specimen of the amalgamation of the two styles i.e. Deccan Daulatabad and Mughals can be seen at Aurangabad and Hyderabad and other places of Deccan. The first Mughal Hammam was constructed during the reign of Shahenshah Babar at Rambagh Agra.[110]

Baths in Aurangabad are either for royal utility or public use. The royal baths are more imposing and are situated within the fort premises, mostly during Tughlaq's period. The total number of Hammans which can still be seen either in good condition or bad condition are six in number. The following classification is according to their condition:

1) Hammam Shahi (Royal Bath) near the Tughlaq palace within the fortification was in a good condition.

2) Hammam near Sunheri Mahal.

3) Hammam near Chand Minar is in a fairly good condition.

4) The public Hammam which is located within the outermost fortification wall near Dargah Shaikh Chand Baoli.

5) There are two completely damaged Hammams

    1) Hammam Dargah Bahauddin Ansari (totally damaged).

    2) Hammam near Tughlaq Palace is in a extremely ruined condition. It contains beautiful decorated stucco plaster and paintings.

Shahi Hammam :

is the most complex Hammam within the fort around 700 ft. away on north east side of Chand Minar. It is built in brick and mortar. The complex has a labyrinth of chambers, cells and apartments. The main entrance is towards the south through steps and a narrow passage which measures 5ft. 3 inch in width and 7 ft, 6 inch in length and opens into a verstibule which is square in shape and measures 18 ft.10 inch x 18 ft.10 inch. The western side of this chamber is octagonal in shape. In its centre a square tank is constructed measuring 76ft.8" in width and length and 5ft deep underground pipelines are attached to the tank. It has a domed ceiling and the room is decorated in stucco plaster and paintings. A round bow at the apex is provided at the vaulted roof to allow light to sink in and discharge bad air.

After the first chamber is a passage in which each room of the bath is connected. The second chamber is slightly smaller then the first chamber, again a square with simple stucco plaster and has been identified with apodyterium by Ansari,[111] a place where the bather undressed and left his clothes. A small water tank is also attached in this chamber. The passage leads to a third chamber which is quadrangular in shape, and still smaller than the first two chambers and is identified with alpiteriun, where the bather used oils, which was anointed to the body. A small corner opposite to chamber three was used as a toilet and measures 12 ft × 10 ft. 6 inch. A small tank has been built here. The holes open in the moat. Typical toilet seats are visible and reveal more of Roman influences.

The northern side door of the passage opened into chamber four which is one of the biggest chambers measuring 14ft 3 inch x 16ft 3 inch. One feet above the pavement of the chamber is provided with platforms or cells on its east and west. In its centre is a tank and its depth is 5ft., with underground pipelines attached to it. The chamber had multiple uses. The eastern cell was used for hot water bath. The middle chamber was identified as calidarium or hot room. The last north corner room is a water reservoir used for number of purposes for heating water as well as a bath tub. The ceiling of this chamber is also domed. Ventilators are provided in the vaulted dome. Another chamber to the north 9ft x 9ft was used for water heating purposes.

The outer walls of this hammam are very thick and measure 4ft to 8ft in thickness while the inner walls are 2ft 6 inches in its thickness. The water was conducted from one room to another by earthern pipes through channels. The pipes are ½" thick and 4" inch diameter. The pipes, the ground plan, each chamber and its use were all well planned and designed. This is one of the most complex hammams found within the precincts of Daulatabad fort and the entire design is mind boggling as one enters each chamber and witnesses the well-laid tanks, water channels, pipes, ventilators and beautifully laid pipelines (Plate XXV).

Public Bath :

Located near Dargah Shaikh Chand Baoli. Presently the location of this Hammam is opposite the fort across the road behind the row of eating outlets and not visible in the first instance. It is towards the east. This is one of the largest and most magnificent hammams in the Deccan. This Hamam as can be minutely discerned was originally Muhammed Bin Tughlaq's construction but during each period underwent not simply renovations and repairs but also alterations in the original designs especially during the period of Malik Ambar (Nizam Shahi of Ahmednagar), and Aurangzeb. This particular Hamam has been described as Ghusal-Khana in Yadgar-i-Daulatabad (Plate XXIV).[112]

The entire hammam measures 97ft in length from east to west and 53ft from north to south. The actual height of the hammam is 13ft 9 inch except the front portion which has a upper storey which probably is a later construction. Most of the construction is in brick and mortar. Wall is as thick as 5ft to 6ft. The rest of the construction is in brick and mortar.

There is only one entrance which is provided with a chamber including two cells. Both the sides of the chambers are open cells with arches. The roof of the cell is in arch style. In the north corner is a water tank with an air tower attached to it and it is through this small tank and air tower that the pipeline enters into the hammam. The chamber measures 13ft 6 inch.

Main entrance has a front portion with three cells measuring 58ft. in length and 13ft and 6 inch in width and is provided with an upper storey. It consists of four chambers which are in a ruined condition. The first chamber is square in plan and supported by four huge pillars. It has a central water tank. The pillars have stucco plaster which are highly decorated and designed. The walls consist of niches. It has a domed ceiling and is identified with apodyterium, as well as lounge.

This chamber opens into a narrow passage measuring 25 ft.6 inch in length and 5 ft 3inch in width. The south corner has a toilet and north corner a narrow passage occupied by a small chamber with miniature water tank. From north corner one goes into the second biggest chamber measuring 48ft x 48ft. with a labyrinth of chambers, cells, apartments and tanks.

Chamber A, B, C, D, E and F are all square in plan and equal in size approximately 13ft x 13ft and 8ft 3 8ft 3 inch The chambers ABCD were used for hot as well as cold water bath are E and F for tub-bath. Chamber G and H are also similar in plan and semi-octagonal and are provided with niches and miniature water tank.

All these chambers are interconnected with channels and earthen pipes, and are systemetically embedded in masonary, moving in horizontal and vertical fashion. The ceiling of these chambers are mostly vaulted or domed covered with stucco plaster work. The vaulted and domed ceiling has square, rectangular lotus designs. The stucco plaster and lotus designs are aesthetically beautiful and appear pleasant to the eye.

In short the entire hammam is a beautiful well-planned complex arrangement of octagonal, semi-octagonal, rectangular and square shaped chambers of varying sizes, connected through narrow passages, corridors. The underground earthen pipes which inter-connect from chamber to chamber have been laid down with the original construction.

An inscription at the main entrance, a text from the Quran, the first line of which is in Arabic mentions, 'Allah's wish is but to remove uncleanliness far from vow O folks of House hold and clean with a thorough cleaning. A persian inscription mentions the name of Muhib-Jan (lover of the Holy prophets family), and a calligraphist and engraver Abdul Qadir Hussain. The date of the construction according to the inscription is 1582 A.D.[113]

Hammams and Its Utility Value :

Hammams prima facie utility value was to provide the facility of physiotherapy clinic.[114] It also means to bath (Hammam) and the term gusl khana (persian) means a bathing place.[115]

In Medieval period mass scale travelling was involved as security or permanent abode in one place was near impossible especially for soldiers who existed in huge numbers. They required not simply resting places while travelling, but also cleaning the body, physiotherapist treatment due to extreme body fatigue due to continuous travelling. Hence these bathing places provided the most soothing,relaxing comfort to a tired traveller.

Technical names were given to each kind of treatment. The first was called as apodyterium,[116] where a bather was supposed to remove his clothes while the second stage various pure and medicated oils were rubbed on his body and is called as Alipterium.[117] Provision of medicated poultice was also made available to put on the affected part of the body and the pain like rheumatic pain, kidney pain or liver pain were killed.[118] After this stage the bather enters a hot room called calidarium,where temperature was increased.[119] To cure rheumatic pain, inflamation of stomach and jaundice a treatment of fomentation was used, and consisted in giving heat to a particular part of the body.[120] The temperature of calidarium was raised from 105°F to 125°F.[121]

These baths also consisted of various types of massages like stroking massage, friction, pretrissage, striking massage and massage through shaking and vibration.[122] Later this was followed by a hot water bath, steambath was the main characteristic of Hammam at Daulatabad and these arrangements were made in Sudatoricum.[123] Medicated hot water bath was commonly given to cure a number of diseases. Just like hot water bath, steam bath, cold water bath was also a part of the treatment. The last procedure was the tepidrium which was so called because of a warm room temperature of tepidarium chamber was kept at 80°F[124] so as to help the bather to adjust to the temperature outside. During the Mughal period the tepidrium became more of a meeting place and the Shahi Hammam had more than one uses as most high level secret meetings were arranged in these chambers.[125] It becomes more popular as a private council chamber.[126] The confidential meetings were conducted in these Hammams.

Utility of Hammams at Daulatabad :

At Daulatabad the hammams were used after a payment of fees as recorded by Shahnawaz Khan the author of "Masir-ul-Umra".[127] According to Ansari,[128] the hammam established was separately maintained under a darogha, who was its administrative officer with a vast array of subordinate staff during Mughal period.

Besides the guards and other security officials, it would have comprised of a number of engineers and water experts, who operated the water devices of hammam, light keeper, secretaries, munshis and record keeper and number of menial staff.[129]

Palaces within the Fort of Daulatabad :

Since Daulatabad Fort has been a witness to the establishment of various dynasties, powerful rulers starting from the reign of Yadavas, Alauddin Khiljis, Muhammad Tughlaq, Bahamanis, Nazam Shah of Ahemanagar, Mughals, Marathas and the Nizam of Hyderabad it witnessed large scale building activity. Some of these buildings were of mediocre. Quality while some of the constructions displayed unique characteristics. Again some of these rulers must have spent fairly long period within the fort. Residential requirement naturally demanded palatial residences for the royalty as well as the common man and nobility. The residential houses of the common man where within the Ambar kot. Excavations were conducted by the Department of History. Dr. Babasaheb Ambedkar Marathwada University under the expertise of the renowned scholar Dr. Ramesh S. Gupte in collaboration with Deccan college under the expertise of Dr. Dhavalikar and Dr. Mate between 1972 and 1973 and later continued during the year 1982 and later. A number of remains and ground plans of houses were discovered which revealed the elaborate nature of these constructions. These houses also included underground storages and seperate toilet blocks. In the interiors of Maha kot there are some blocks and houses for common man as well as residents for nobility. But all these are plans which can be discovered only after proper archaeological excavation. But otherwise none of the outer signs are visible.

However some of the constructions of royalty are still visible though is in a very bad state of ruins. The palaces right from the times of Yadavas to Nizam Shahs, Mughals and Bahaman's are still visible. It is sometimes difficult to identify the periods of these palaces and ownership due to total lack of references except for few records available like Nath R., Raunaq Ali, Campbell, Bilgrami, Yazdani, Riyaz Ansari has done extensive and exhaustive work in his thesis and has included ground plans of these palaces in his thesis which are very helpful for this work.

Most of the palaces are in a very bad state of ruin. The ASI under whose control and responsibility lie the palaces seem to have neglected the maintenance of the palaces and allowed them to deteriorate and degenerate at their own comfortable space. Unless urgent and immediate steps are not taken each day a stone will tumble down leisurely and within a few years we will be deprived of our proud heritage, which still stands to narrate the historical events and memories.

The beautiful arches, niches, towering walls of the palaces, intricate designs, decorative motifs etc. will become history if ASI does not take its first and foremost responsibility of preserving, restoring and repairing the historical memories. If

Daulatabad Fort is to be identified as a world heritage monument by UNESCO the accountability of ASI will doubly increase. The proud and romantic palatial remains must be protected for posterity.

The palaces at Daulatabad have their own charm and beauty and its situation on a hill fort lends it added romance and magnificence. But these palaces can in no way be compared to the magnitude, massivences and splendour of the Mughal palaces at Delhi, Agra, and the Rajput palaces at Jaipur, Udaipur and other cities of Rajasthan. The palaces built by various rulers served them more as transitory abodes rather than a protracted abode. Naturally due to their flitting lives and continuous movement none of these rulers spend generously on these residences.

Diwan-e-Am Or Amkhas Palace :

This is the first palace in Daulatabad that the tourist would come across as he turns towards the right pathway passing from the Chand Minar towards the north. According to Raunaq, the palace was used as a meeting place for the king with his umras ard high rank officers.[130] According to Riyaz Ansari he indicates that there is a possibility that this was not a palace but originally a gate later converted into a palace.[131] Local tradition says the palace was used as complaint 'Kacheri'.

Amkhas palace is a double storeyed building measuring 52ft x 42ft and consists of a lofty central arch that rises upto the height of the building. It is constructed in ogee arch or true arch style. The porch or the central hall of the palace has a huge dome ceiling and diamond cut moulding in the interior portion. The central chamber has arcades on either side with similar diamond cut moulding A flight of steps to the north lead to the upper rooms of the palace. The domed ceiling is covered with polished stucco plaster. Towards the southerns corner a flight of steps leads to the roof of the building. The entrance was converted into a palace for two fold reasons for royal residences and council hall. On the west of the building was another gateway that has now fallen down. It served as an entrance to another palace complex and hamam now in a very ruined condition. After crossing this courtyard the other palace complex and hammam are visible (Plate XXVI).

Pavilions :

Location has been selected with extreme care taking into consideration the extremely unstable conditions and unpredictable situations that could arise out of sudden revolt amongst its own soldiers. So far avoiding any such risks pavilions were built on high brinks of the plateau that commanded a view of the low land. Defence is ensured by a deep ditch from the west and a massive wall from the east, measuring 6 mtrs in height while the thickness of the wall is 1.1O mtrs.

In front of the palace is an expansive courtyard very well planned and designed, replete with medieval features adopted for environmental, structural and decorative harmony. The courtyard is quadrangular and has pavilions on both western and southern sides (Plate XXVII).

The pavilion on the western side is well ventilated and is a three chambered pavilion. The main inner octagonal chamber is flanked by two small well ventilated octagonal chambers on either side. There is a square open chamber in front of this with five cushed arches, which are still intact.[132] This has a strong aesthetic quality and are mostly uniform in size and shape. The central inner chamber too is octagonal, it is round surrounded by a low dome ceiling with stucco design.[133] This is supported by eight columns of stone masonary and eight open screens that are surrounded by wooden lintels. The main octagonal chamber is surrounded by a semi-octagonal wall with arcades of arches and with openings. There are total 13 arches surrounding this octagonal chamber.[134]

The main octagonal chamber is flanked by two small octagonal chambers with low domed vaulted ceiling supported by eight columns of a masonary stone and has eight open screens, surrounded by wooden lintels. The columns are richly decorated by stucco flora and stylized designs. There are two small quadrangular rooms and western side of these small octagonal flanking chambers. These small chambers have flights of steps leading to the roof of the pavilion. The roof is flat and octagonal in shape. In front of the main octagonal chamber there is an open court in the middle of which is an open court with a sunk basin in the pavement. A beautiful model of indoor water device.

The basin was earlier decorated with yellow and blue persian glazed tiles arranged in geometrical designs. There are still remains of this decoration.

The open rectangular court has five cusped arches facing towards the east. The facade is much damaged but the arches are intact, and show strength combined with beauty in the style of architecture. It is covered with stucco plaster and cusped (kangura) arches that have floral and stylized design.[135]

The pavilion on the south side is in a bad state of preservation as the ceiling has collapsed. It appears that the ASI is least bothered about repairs of such beautiful palace as the fragment remains lie in a state of ruins and the area is crowded with growth of wild bushes and trees. However the remains are sufficient to show that the pavilion had three chambers with open court, and floral designs on the wall.

The main chamber has two chambers similar in design and shape. Unfortunately the roof here also has fallen down. The wall has niches all round and stucco floral and stylized design. These chambers have two cusped arches covered with stucco. The ruins are still scattered around.[136]

On either sides of this pavilion there are two chambers probably accommodation for guards according to Ansari.[137] This chamber is also in ruins as its roof has fallen down while the masonary walls are still intact. Due to the major ruined conditions its original plan is difficult to recognize. The fallen material still lies down.

Am Khas Garden :

During the medieval period and especially during the Mughal period gardening was always associated with architecture. The Mughals laid great emphasis on this aspect. Babar introduced various devices to provide adequate running water in and around the building to bring relief from the burning heat of Agra.[138] The area in front of these two royal pavilions is occupied by a garden rectangular in plan. It measures 47.75 mtrs. (north to south) and 43.53 mtrs (east and west). It is a beautiful example of an indoor garden laid out in geometrical shape at this complex.

This garden originally was square in shape, but later two more rectangular plots were added due to which the shape underwent change. Though all the features like tanks, channels, paved walks in lime and stone are laid down, the garden does not possess the same beauty like the Mughal garden. But standing individually minus comparisons it is extremely pleasant to the eye despite its ruined conditions.

There is a central tank and the whole garden is divided into six plots and the tank is provided with inlet and outlet. The tank is one meter high from the ground level, provided with three steps from each side. The four plots of the garden ABCD are equal in its length and width, the remaining E and F are smaller.[139] The pavement leading to the tank from all direction varies in its width.

Underground pipes well connected to extraordinary strength and resilience provided water to the tanks, fountain and basin. Just as in Agra inside the palace at Fatehpur Sikri there were underground pipelines to cool down the temperature, similarly the garden of Amkhas also ensures regular supply of water to cool down the Deccan heat during summers. The present state of the pipelines is far from satisfactory as most of them are dead and do not serve the original purpose of maintaining the garden.

Chini Mahal :

Also known with this name due to the glazed tiles decoration on its facade. Situated near the eighth gateway of the fort it is completely inlaid on its face with blue and yellow glazed tiles (Plate XXVIIIA). Glazing is generally prepared by a process in which colours are pasted first and are coated over with the glaze and than the whole is fused in this process the glaze retains its identity over the colour.[140] The height of this beautiful palace is about 10 mtrs and is 15 mtrs in length and 5.80 mtrs in width. This is a double storied building and appears very impressive despite its present ruined condition. The roof has literally collapsed but its walls are still intact.

Unfortunately if urgent renovations or repairs are not undertaken the building will definitely collapse. Cracks can be witnessed in its side walls and bricks falling in gradual numbers and can be seen strewn around.

The ground floor is divided into two section a higher and lower one. The first section measures 15 mtrs x 2.80 mtrs and the upper section measures 15 mtrs x 3 mtrs in length and width. This floors, roof has fallen down. A flight of steps are visible from the eastern side of the facade, leading to the first floor.

The upper floor is well ventilated and has two chambers. The inner chamber measures 2.80 mtrs x 5.80 mtrs and has a small door on the east and niches in the wall. Another chamber measures 7.80 mtrs x 5.8 mtrs and has three painted arches on eastern and western wall of the palace. The roof of this floor too has collapsed.

Sultan Abul Hasan Tanashah's association with the Chini palace :

The Chini palace since decades and centuries has been closely associated with the king of Golconda. Abul Hasan Tanashah was confined in the palace and imprisoned by Aurangzeb in 1687 AD and died in the year 1700 A.D.[141] His stay extended to 14 long long years. Bilgrami who has quoted Meadows Taylor,[142] a British officer who writes about Tanashah as follows. 'A minstrel himself, his ballads and amatory poems have survived him and the regret at his disposition was deeply and universally felt by his subjects. Golconda fell in September 1687 A.D., hardly year after Bijapur and the king was sent a prisoner to the fort of Daulatabad where he resided till his death, 14 years afterward and where as in the Deccan generally, many anecdotes of his as the good and brave king are still current."

Sultan Abul Hasan Tana Shah was the third son-in-law of Sultan Abdullah Qutub Shah, who ascended the throne in 1672 AD and ruled for 14 years. The last seige of Golkonda by Emperor Aurangzeb marked the extinction of the famous Qutub Shahi dynesty to the Mughal Emperor and the once glorious reign of Hasan Tana Shah was ended. After this period he was held captive till his death.

Hasan Tana Shah was buried in the dargah of Hazrat Sayed Shah Raju Qattal Hussaini, at Khuldabad as he was his mureed or disciple. The story in Abdul, Haisbook Gulistan-e-Khuldabad,[143] runs as follows. Once the kiledar of the fort in his dream was instructed that Abul Hasan Tana Shah's body should be buried in the premises of Shah Raju Qattal's tomb. The kiledar in the morning sent a message to Tana Shah requesting him for an audience. Tana Shah refused saying that he is aware of the reason for the audience and the meeting is not necessary. The Kiledar however should follow the instructions of his dream, and act accordingly. Thus Abul Tana Shah was buried as per the instructions. And his body lies in the same premises.

Though Abul Hasan Tana Shah did not directly belong to the royal Qutb Shahi dynasty was the son-in-law of the last Qutb Shahi ruler through succession and was fortunate enough to reign for fourteen long years. The number fourteen had a remarkable influence in his life according to Raunaq Ali.[144] Abul Hasa Tana Shah spent fourteen years in boyhood, remained a devout disciple of Sayed Shah Raju Qattal, ruled for fourteen years and remained a prisoner inside the fort of Daulatabad for fourteen years.

The palace still retains the glazed tiles, though few in number (Plate XXVIIIB). Similar glazed tiles can be viewed on the Chand Minar inside the fort. The construction of the building has parallels with the Bahamani buildings of Ahmed Shah II or III. Similar building can also be seen in Bidar and Bijapur, like Chini Mahal, Nagina Mahal, Gagan Mahal that represent the Zenith of Bahamani art and architecture.[145]

Sonheri Mahal :

The palace is situated as one walks inside the eighth gate, towards the left and the Mendha Tofh Burj towards right situated on an elevated location. The palace does not retain any of its golden colour, unlike, the Sonheri Mahal located at Aurangabad which still has the most intricate delicate floral designs painted on the inner enclosure of its ceilings and the wall and parts of the pillars. Probably the golden paintings disappeared due to the complete exposure as the major portion of the roof has collapsed lending it little protection from all climatic conditions through so many years.

The Sonheri Mahal at the entrance has a huge courtyard with raised platforms on either sides. The facade of the palace contains a number of niches. The total dimensions are 23.85 mtrs (east to west) and 14.5 mtrs (south to north) the height of the palace is 5 mtrs. The whole palace consists of 6 chambers connected with each other. In most parts the roof of the palace has given way though the inner corners of the roof are still visible.

The entrance doorway has a huge arched opening with floral motif on either sides above the arch, there are slim, slender columned pilasters on the lower sides (Plate XXIX). The stucco plaster is still visible. The span of the arch is 5.50 mtrs and height is about 4mtrs upto parapet. The lower portion of the arch has wooden lintels. The parapet has merlons. Both the interior and the exterior portion of the palace consists of several niches in the wall, identical to that of Barid Shahi Palaces of Bidar.[146]

The palace is on a lower land then the Rang Mahal on the back side and consists of six chambers interconnected (Plate XXX). The two inner corner chambers have upper storey and the remains of these stories are still visible. The upper storey is approached by the flight of steps which are rebuilt on the eastern end of the building.

The centre chamber has two chambers on either side. All the inner chambers are uniform in both style and size. The opening to all the inner chambers have arched openings. The central bigger chamber leading to inner chamber has three arched opening on either side. The corner chambers still retain the ceiling and the ceilings has a wooden support, wooden pilasters with lintels with beautiful designs and decoration. At Daulatabad this is the only palace with wooden work (Plate XXXI).

The upper chambers have conical domed ceiling. These types of conical domes were constructed on an anonymous tomb near Dargah Bahauddin Ansari and main

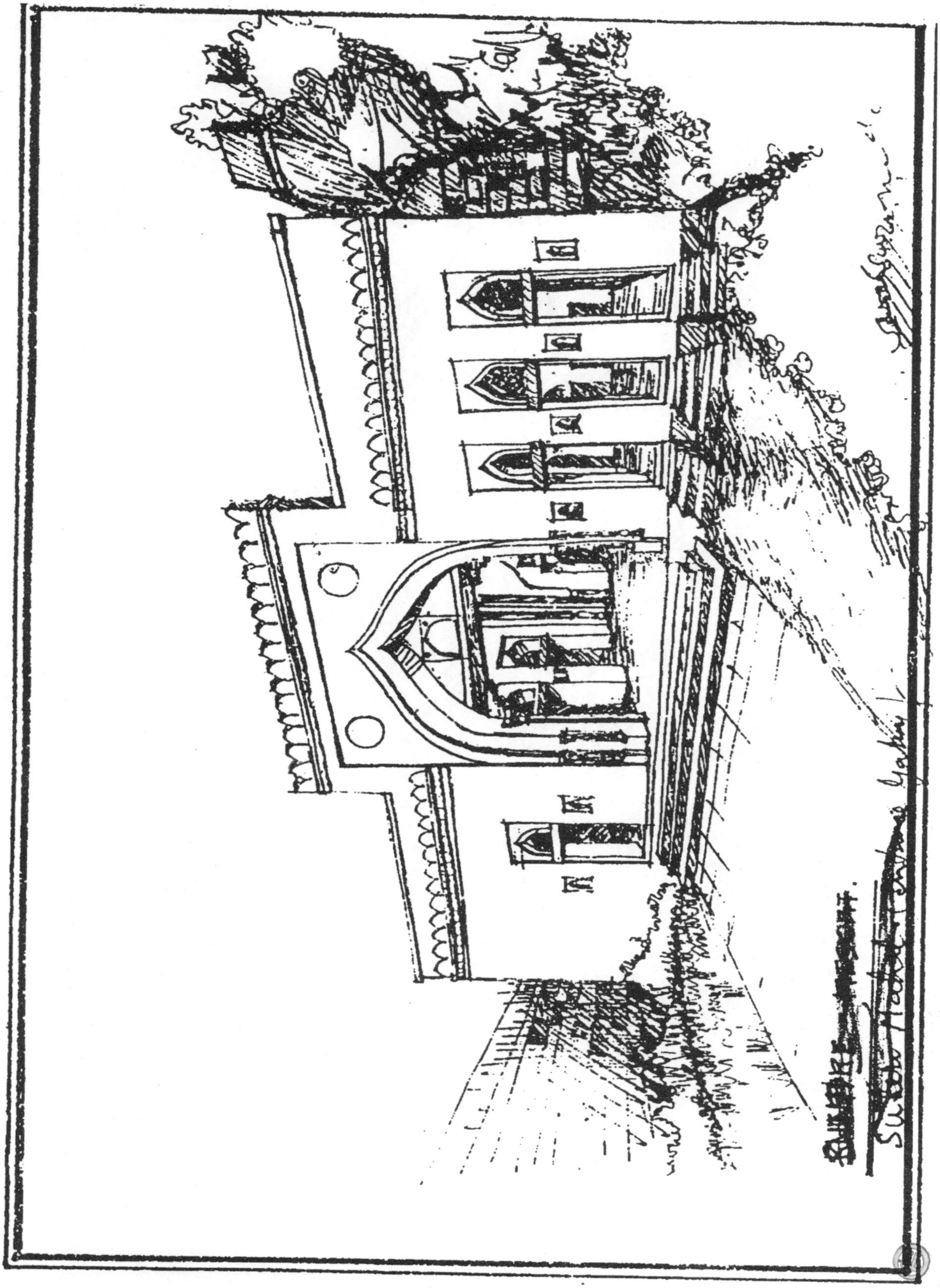

Fig. XI. Sunheri Mahal (Entrance Gateway).

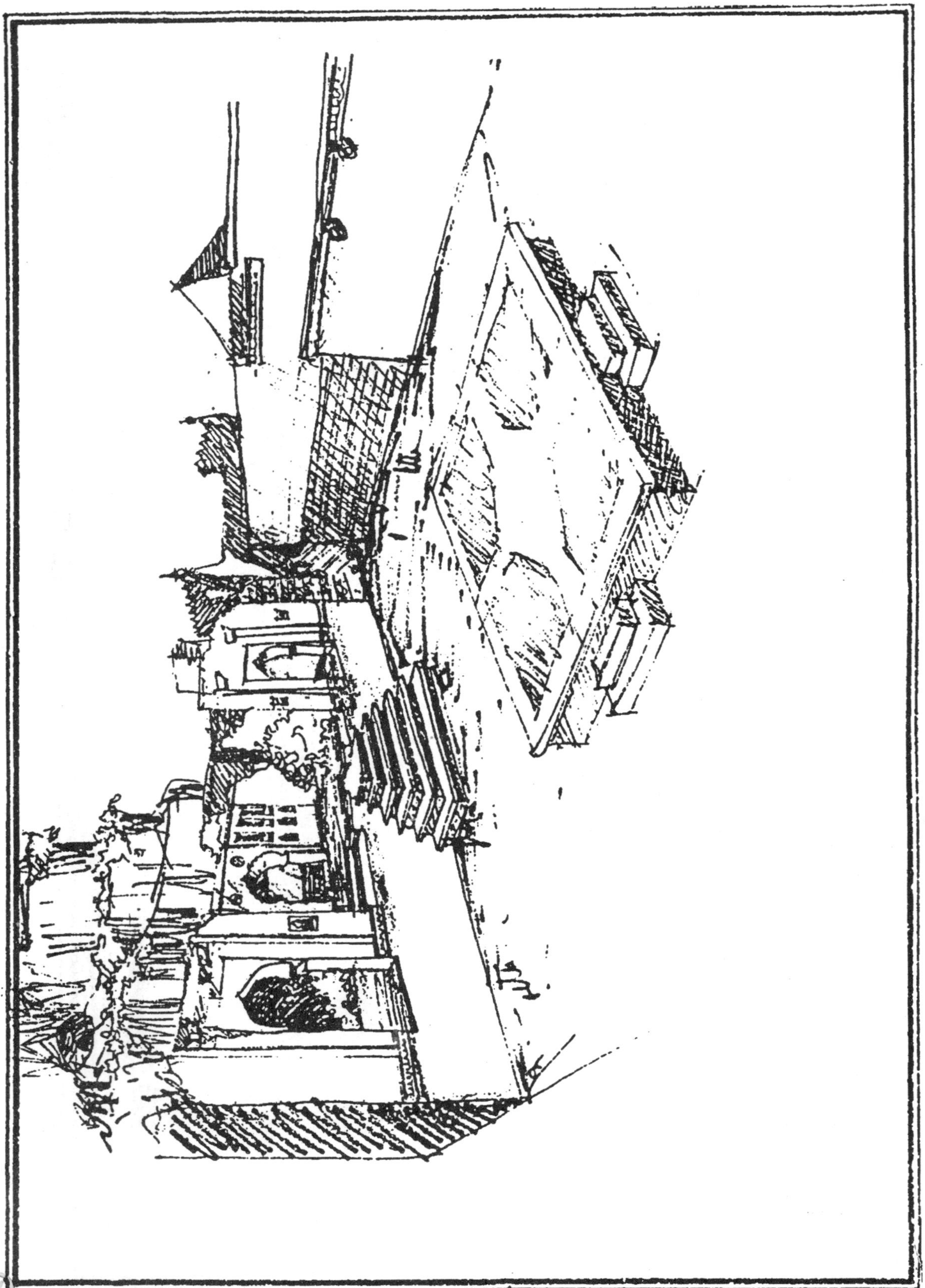

Fig. XII. Rang Mahal.

tomb of Momin Arif at Daulatabad.[147] Similar domes are found among the buildings and tombs of Bidar such as the tomb of Waliullah and Kalimullah Shah Bahmans.[148]

The entire palace was covered with stucco on its interior as well as exterior. Both brick and stone is used in the construction. Stone rubbles are seen scattered around and there is wild growth of grass in the corners and above the walls. The walls are 85 cms in thickness and are supported by wooden lintels, columns and beams all of which are most artistically carved. The palace is the only exclusive example of carved wooden work, now in a very bad shape on the verge of crumbling.

Rang Mahal :

Entrance to this palace is from the seventh gateway of the fort and the palace is situated towards the south side of Sonheri Mahal. The palace has three open entrances of which the central entrance has fallen down (Plate XXXII). This was one of the best specimens of Mughal architecture and displays some of the most artistic decorative motifs like cusped arches, miniature domes, minarets and pillars. Ornamental designs are visible at the entrance. Between the entrance and the palace is a paved courtyard and has a chabutra built in stone. The chabutra is in a bad shape with wide growth of grass and the stones of the chabutra are gradually falling apart. The entire palace is built on a high platform facing to the east. After crossing a flight of five steps we can reach the middle chamber of the palace. The total inner dimensions of the palace measure 35.70 mtrs in length and 9.5mtrs in width. [149] The height of the palace is 3 mtrs. Complete roof of the palace has collapsed but the walls are intact and exhibits several small niches decorated with stucco plaster as was visible in Sonheri Mahal too.

The rooms too are decorated with stucco plaster. The interior arches are engrailed and is decorated with stucco, floral designs and medallions. This type of stucco plaster has been used on the buildings of Bijapur and Bidar, such as Gogumbad and various mosques and palaces of Bijapur and Bidar.[150]

Shah Jahan's Baradari :

The Baradari a white structure visible from any angle is the most conspicuous palatial construction attributed to Shah Jahan and is said to have been the favourite summer residence of Emperor Shah Jahan and his son Alamgir Aurangzeb.[151] Raunaq Ali reports that this was constructed on the instance of Shah Jahan in 1010 hijri.[152] Encyclopedia of World Art also records that the Baradari of Daulatabad was constructed during the reign of Shah Jahan in the year 1636 A.D.[153] (Plate XXXIII).

The white octagonal building is situated near the summit of the fort and has access from Fakir Sultan's Shrine. Adjoining to the Baradari is a square courtyard with open verandas and galleries all round. The entrance is towards the southside of the building. Towards the north of the chamber is a toilet block. A double storeyed building, the visitor enters it from the south into the pavilion. The main courtyard measures 50ft x 50ft. In the centre is a platform above 1 ft 4 inches from the ground level, measuring 12 ft 2 inch, east west and 9ft to south-north. The northern and southern chambers of the Baradari are uniform in its style of architecture. Through the southern chamber a visitor can enter into the courtyard of Baradari and northern chamber leads to Hathi tank and Moti tank.

Baradar is mainly constructed in black chiselled stones of uniform size. The eastern facade portion of Baradari is octagonal in shape and has two octagonal chambers on either side. The southern main courtyard measures 50 ft in length and 17 ft in width. This courtyard has two octagonal chambers on either side of uniform size. Each octagon has three open arches which are pointed supported by masonary pillars of quadrangular shape. Above the arches is cornice of stone supported by simple brackets. The brackets are placed at uniform distance over the pillars of arches. The upper portion of the baradari has parapet of simple kangura style. The roof of the galleries is dome shaped (Plate XXXIV).

Sas Aur Bahu Ke Makan (House of Mother-in-law and daughter in-law :)

According to Raunaq Ali the two houses to the south of Kauri tank are known as Sas aur Bahu ke Makan.[154] Even the local traditions also record the two houses as Sas aur Bahu ke Makan. Not much is known about the origin of the houses or the exact reasons for naming them as mother-in-law and daughter in-law's houses. The first house measure 76 ft in length and 28 ft in width and is supported by five pillar with a courtyard measuring 76 ft x 28 ft. Second house is much smaller and measures just 38 X 37 ft. south west.

The topmost portion of the fort proudly displays a huge canon 64 ft in width and 80 ft in length and 28 ft in height with a marathi inscription,[155] stating that it was made from the funds for casting provided by Mangaldas Krishnadas Lavji and Raghavaji the Lavji Banias of Gujarat. A persian inscription states that the gun is named as 'the creator of the storms' Durga Toph.

The Tanks Moti and Hathi Tank, Kauri Tank :

The Tanks are also situated near the Baradari. The Moti tank is located to the north of Baradari with a flight of steps leading to it. It has a room of two chambers excavated into solid rock of the fort. The tank receives natural spring water and no aqueduct or artificial water arrangements. It was probably the handiwork of the Yadavas. Below the Moti tank is Hathi tank, measuring 40 ft x 30 ft. It has also natural springs. During the monsoon the water overflows here. Little beyond this is a third tank called Kauri tank also excavated in solid stone with two room chamber. This tank too has natural springs. Even in summers it has natural springs.

The Kauri tank is a natural device in the rock filled with spring water. It is beyond the above mentioned tanks situated 30 ft to 35 ft from Baradari towards the west.

Janardhan Swami's Samadhi :

Janardhan Swami was a renowned saint of Maharashtra and the pride of the state whose teachings spread all over Maharashtra in the 16th Century. He was more famous as the mentor of Eknath Maharaj and held in high esteem. Maharaj Janardhan was also the prime minister and chief advisor to the Governor of Daulatabad and also a Killedar of the fort for a short duration. He put most of his thoughts in writings few of which exist today. His shrine is located on the right of Bastion Bala Hisar or the summit where a small winding pathway leads to this excavated cave used as a residence by the Janardhan Swamiji. Pilgrims throng this small cave during Margashirsha Vadya and a fair is held in his honour for three days.[156]

Chand Minar :

A minaret visible from miles attract attention of all tourists, is a Tower of victory", built by Sultan Alauddin Bahman Shah. The minar does not have any functional value but is more a decorative piece of architecture, which had its earliest precedence in the minaret built on the orders of Abdul Malik in 685 A.D. Abdul Malik was appointed incharge by the Governor, Oman-bin Abdul Aziz. The next minarets were seen at the great mosque of Damascus, begun by Abdul Malik in 705 A.D. and finished by his son, Walid I 705-715 AD.[157]

The landmark of Indian minarets is Qutb Minar also a tower of victory planned by Qutbuddin Aibak in 1197 but completed by Iltutmish who ruled between 1211—1236AD. The Qutb Minar was originally 225 ft in height tapering as it went up and consisted of four storeys essentially, Islamic in form and design.[158] Later the Qutb Minar was struck by lightning and its fourth storey was damaged. But Firuz Tughlaq (1351-1388) replaced the upper storey by two smaller ones and raised the height of the tower to 240 ft.[159] Another famous Minaret in India is the Char Minar of Hyderabad with four minarets at each angle. However it lacks both in height and grandeur compared to Qutb Minar in Delhi and Chand Minar at Daulatabad Fort.

Chand Minar not only compares in height (210 ft)[160] with Qutb Minar but also in its glory and Grandeur. The minar is surrounded by spires and was erected in 1435 A.D. by Alauddin Bahman Shah, one of the most majestic, eye catching architectural feat of the Bahaman dynasty. The ruler of Deccan during this period was Ahmed Shah Bahman-II celebrated ruler for his architectural miracles. Ahmed Shah Bahman-II is also responsible for transferring his capital from Gulbarga to Bidar in 1425 A.D. Thus giving impetus to his building skills and talents.

A persian inscription within a small mosque attached to the minar described sultan Ahmed Shah Bahamani as brilliant as the moon and who was so highly wise and intelligent that it could put to shame even the most famous king Jamshed of Iran once famous for his wisdom. On a certain day the king called Alauddin—

Bahamani into his presence and presented him the fort of Daulatabad. Alauddin created this huge grand Minar a 'Tower of victory'.[161]

The Chand Minar was built in 1435 AD to commemorate his conquest of the fort of Daulatabad. Another account says that the minaret was erected in commemoration of the first conquest of the place by Muslims in the year 1294 after the invasion of Alauddin-Khilji who conquered the fort in the year 1294 when the Yadavas King Ramchandra was the king of Daulatabad[162] (Plate XXXV-XXXVI).

Such a possibility is remote as Alauddin Khilji is simply said to have conquered the fort, surrendered and looted its wealth and left with the plunder towards the north.

The impact of Iran is evident not only in the Qutb Minar architecture but also the Chand Minar which probably was inspired by Qutb Minar. The first minaret to be built in the Persian style though it was also influenced by indigeneous features such as the carved ornamental brackets supportive to the balcony and other details as tiles, cylindrical shape sloping from bottom to top, but in design it is different from the other minarets (Plate XXXVIII).

According to Ansari the minar measures 210 ft and its circumference at the basement is seventy feet. The tower is erected on an oblong platform which is about fifteen feet in height constructed of solid blocks of stone. The entire minar is made in pieces of stone joined by lime, the most powerful binding material popular in medival period. In some areas where the blocks of stone have given way to irregular pieces of stones and simply fixed without any binding medium.[163]

At the base several chambers as many as 24 chambers are evident with extremely austere architecture, and no decorative motifs either within the chambers or outside the chambers. The exact purpose of these chambers is not clear. They are all arched shaped chambers (Plate XXXVII).

Towards the north of the tower there is a Diwan-Khana containing three chambers supported by two pillars which are square in shape. The pillars are devoid of any carving and are very simple, with a window in the centre. The exterior portion of this window is actually a flanking balcony which is supported by ornamental brackets which are four in number. The interior chambers of the Diwan Khana are domed shaped while from above the roof is flat.

On the South of the minar is a small, but a beautiful mosque, a rare piece of architectural genius of Indian Craftsmanship and creative ingenuity and fitness of the Sultan Alauddin Bahaman Shah. The wall is of the mosque contain niches with a mihrab in the centre but no pulpit. On the south wallis a large tablet containing a persian inscription dated 849 High/1435 AD as discussed earlier. On the South of the mosque is a flanking balcony, which is supported by ornamental stone brackets which are four in number and serving the purpose of ornamentation.

Chand Minar tapers towards the top as to maintain balance and reduces heaviness as it moves upwards. It has three galleries which skirts the exterior at various regular heights. On the roof of the mosque from the south side it was entered by a doorway within which a stairway spirals it, way upto each balcony finishing as the platform on the uppermost that means the third storey of the minaret, and in its artistic aspect the most elegant aspects of the minaret are the balconies and designs (Plate XXXVIII).

The lowest balcony stands about sixty feet from the ground and is supported by artistic ornamental stone brackets. These brackets are thirty three in number and most of them are in good condition. They contain decorative bands. The gallery has a balustrade all around. These balconies took the form of stepped battlements or merlons called Kanguras a fragment of which showing the pattern still survives.[164] The ancient ancestry of this distinctive motifs may be realized when it figures in the architectural motifs of Assyrian palace of Sargon, Khorasabad even more original is a system of stalactite bracketing underneath the balconies by which these projections are supported and their weight transmitted to the body of the structure.[165] No doubt these brackets which supported the gallery or balconies are indigenous in character but the idea is taken from Persia.

The second balcony also possesses balustrade all around with merlons or battlements and also supported by brackets. But these brackets are very simple. The number of brackets reduced to sixteen in number. Beautiful woodwork is used to support the balcony. The construction seems slightly later than the first floor.

As mentioned earlier the minar reduces in size as it gradually tapers upwards. The third floor is slimmer in circumference than the second and the second slimmer than the first. This is also supported by brackets but they are different from the first and second balcony projections. The difference lies in the gap between the merlons in the first and second gallery but there is no gap between the merlons of third gallery.

Added to the brackets, merlons and beautiful bands as decorative motifs are the blue tiles mostly in diamond shape in rows. But unlike the Chini Mahal here the tiles are not genuine tiles but light blue colour cut to diamond cut squares, applied on the chuna or lime coating of the pillar. Though some of the persian genuine tiles are visible from the backside minar and on the upper portion. The Persian tile impact is obviously work of Persian craftsman.

The upper floors can be reached by spiral staircases within the minar most of which are made of solid block of stones firmly connected with each other by lime binding. The summit of this lofty minaret is conically shaped and is surrounded by a crescent, which is about 15 ft in height.

Similar minarets in the Deccan is the minaret of Mahmud Gawan's Madarsa at Bidar erected in 1472 A.D. and at Raichur built in 1514 AD.[166] As evident by the datings both the forts are later than Chand Minar and hence inspired in all probability by Chand Minar, while Chand Minar was an inspiration of Qutb Minar. The Chand Minar from the top offers a beautiful view of the entire area. The fortification walls, the dilapidated palaces, hammams, and mosques are clearly visible. The forlorn minar stands silently year after year, decades after decades, century after century witnessing the process of destruction and ruins.

Second Minar a Memorial to Akbar Badshah :

According to a few stray references, Daulatabad once possessed two minars. The second minar in Daulatabad was a special monument to mark Burhan Nizam Shah of Ahmednagar's debt and high regard for Badshah Akbars assistance in time of crisis. The monolith minar reads range. Akbar's Nishan according to Premchand Kayasth Shayar who lived in Daulatabad. This incident has been recorded vividly in his Shayri or poetry.[167]

Badshah-Namah an illustrated manuscript written by Abdul Hamid Lahori, the court historian of Shah Jahan has illustrated a painting of Daulatabad where both the minars are clearly visible. The exact location of Akbar's minar is difficult to point out but it was smaller in size and lay in front of the Chand Minar. Probably near the entrance. However no traces are presently visible (Plate XXXIX).

Other Architectural Remains within the Fort :

Within the fortified walls of the fort several architectural remains are still evident, scattered here and there. In the very first entrance and second gate there are rooms for guards now occupied by a number of guns and cannons. The guns and cannons are huge in size with beautiful designs which include floral and vine designs, ships with flags within the floral designs heads with horns can be seen. Animal heads are also carved on these cannons.

On the courtyard again towards the left corner is the caretakers office which was a mosque originally.[168] It still appears beautiful though its modern look has slightly ruined its originality.

There is a huge space between the third and fourth gateway. Its left portion which is very spacious has been converted into beautiful lawns and its walls support several stray Jain sculptures from the Jain temples. The balcony of the fourth gateway was once known as Nagarkhana.

As one emerges outside these gateways is a rough stone pathway measuring 1500 ft in length and 12 ft in width.[169] Running between two walls 56 ft high. On the right side of the pathway there are several arched openings probably used as guard rooms or to house animals.

Bandar Khana :

On the right is Bandar Khana surmounted by the ruins of garden and houses.

The garden is known as Rashid Bagh,[170] presently more of a jungle. But recently the ASI has made an effort to clear the jungle and transform it partially into a lawn. A large tract of the jungle has been cleared but is surrounded, on all sides by thick jungle giving it a very cosy appearance in the comfort of thick growth of trees. On the outer side of the garden wall is a set of vaulted chambers nearly seven in number. Most of the Bandar Khana is in a state of destruction. This is a multi apartment building with open chamber and anti chamber with low vaulted domed ceiling. Infront is a Hauz with cisterns connected with underground pipe lines from the Nahar of Aab-Pash-Darra.[171] Reference fig. 73 to 76 fig 77.

REFERENCES

1. (a) Suany Kumara Ananad—Early India Architecture Gifies and City Gates Journal of Eastern Art Vol. 2
   (b) Dutt B.B. Town Planning in Ancient India.
2. Yazdani, p. 23
3. Ibid.
4. Ibn-e-Batuta, p. 227
5. Compbell 'Glimpses of Nizam's Devminian, p. 332
6. Riyaz Ansari 'Medieval Daulatabad Temples, p. 439
7. Ibid.
8. Cambridge History of India, p. 631
9. Ibid.
10. Bilgrami and William Moat, p. 414
11. Stuart Piggot, p. 82
12. Ibid.
13. Riyaz Ansari, p. 443
14. Ibid.
15. Ibid, p. 444
16. Secly John, The Wonders of Ellora, p. 363
17. Raunaq Ali, p. 111
18. Riyanz Ansari, p. 445
19. Ibid., p. 447
20. Ibid., p. 451
21. Ibid., p. 452
22. Raunaq Ali, p. 111
23. Riyaz Ansari, p. 464
24. Ibid., p. 464
25. Ibid., p. 465
26. Ibid.
27. Raunaq Ali, , p. 41
28. Sidney Toy, p. 34
29. Yazdani 'Bidar', p. 32
30. Riyaz Ansari, p. 470
31. Raunaq Ali, p. 48
32. Jairaj Bhog, p. 139
33. Riyaz Ansari, p. 471
34. Ibid, p. 473
35. Map of Daulatabad (Tughlaq period)
36. Sidney Toy, p. 38

37. Lahori Abdul Hameed 'BadshahNama', p. 568
38. Sidney Toy 'Castles', p. 1, 16, 22
39. Ibid., Strongholds of India, p. 3
40. Ibn-e-Batuta, p. 227
41. Bilgrami and William Moat, p. 432
42. Sidney Toy. oficit, p. 38
43. Aurangabad Gazetteer, p. 970
44. Bilgrami and William Moat, p. 432
45. Riyaz Ansari, p. 480, 481
46. Ibid., p. 481
47. Bilgrami and William Moat. op. cit., p. 434
48. Riyaz, Ansari, p. 482
49. Raunaq Ali. MSS, p. 81-82
50. Sidney Toy, History of Fortification, p. 39
51. Riyaz, Ansari, p. 488
52. Raunaq, Ali, p. 111
53. Ibid., p. 104
54. Winter F.E. 'Greek Fortification', p. 152
55. Ibid., p. 206
56. Jairaj Bhog, p. 18, 21
57. Sidney Toy, p. 5
58. Ibid.
59. Winter, F.E., p. 234
60. Raunaq Ali, p. 111
61. Ibid.
62. Nath, R., 'Colour Decoration in Mughal Architecture, p. 66
63. Sidney Toy, 'Castles', p. 90
64. Riyaz, Ansari, p. 484
65. Siendney T. Strongholds, p. 3
66. Ibid. op. cit., p. 3
67. Riyaz, Ansari, p. 485
68. Ibid., p. 404
69. Ibid.
70. Bharatiya Itihas Sanshodhan Mandal, July 1935, p. 19
71. Consens Henry, op. cit., p. 37
72. Siddiqui Saifuddin, op. cit., p. 30
73. JASB, Vo. LXXIII. Part I. Ferishta op. cit., p. 276
    Barani, op. cit., p. 514
74. Percy Brown, op. cit., p. 17 Plate IX
75. Ibid., p. 73. Plate LXIII
76. Yazdani 'Bidar', p. 166
77. Bilgrami and Moat op. cit., p. 329
78. Raunaq Ali, op. cit., p. 1067
79. Bharat Itihas Sanshodhan Mandal, July 1935, p. 79
80. Aurangabad Gazetteer, p. 969
81. Riyaz Ansari, p.421
82. Ibid., p. 422
83. Nath, R.P., p. 22
84. Shukla, D.N. Vastu Shastra. Vol. 140 (Inchnow 1960).
85. Nath, R., op. cit., p. 22
86. Ibid.
87. Riyaz Ansari, p. 424

88. James Burgess. 'The Muhammedan Architecture of Ahmedabad. Part.II, (ASI New Imperial Series Vol. XXXII, London 1903). Plate LXXX, p. 1-2
89. Nath R. op. cit., p. 23
90. Ibid.
91. Riyaz, Ansari, p. 427
92. Ibid., p. 426
93. Ibid.
94. Dr. Syed, Gausuddin, p. 10
95. Riyaz, Ansari, p. 513
96. Encyclopedia Britannica, Vo. 3, p. 204
97. Nath R. Some Aspects. of Mughal Architecture, p. 40
Dr. Syed Gausuddin op. cit., p. 4
98. Ibid., pg. 11
99. Ibid. op. cit., p. 40
100. Ibid.
101. Dr. Syed Gausuddin, op. cit., p. 13
102. Jairaj Bhog, p. 14
103. Ibid., p. 24
104. Ibid.
105. Iqtidar Hussain and Qazi Muhammad Ahmed. Shihabuddin All Umar's Masalik-ul-Ansari (Aligarh 1971), p. 13
106. Sketch of the Hamam.
107. Riyaz Ansari, p. 519
108. Ibid.
109. Yazdani, G., 'Bidar' Its bouy and Manuments, p. 51
110. Nath, R., op. cit., p. 42
111. Riyaz, Ansari, p. 524
112. Raunaq Ali, op. cit., p. 122
113. Riyaz, Ansari, p. 528
114. Dr. Syed Gausuddin, op. cit., p. 1
115. Molvi Noorul Husan Nuggar, 'Nur-ul-Lughat' Vol. 4 (Liveknore 1927), p. 467
116. R. Nath, op. cit., p. 40
117. Plan of Royal Bath
118. Dr. Syed Gausuddin, op. cit., p. 62
119. Ibid.
120. Ibid., p. 47
121. Ibid., p. 2
122. Ibid., op. cit., p. 38, 41
123. Nath, R., op. cit., p. 40
124. Dr. Syed Gausuddin, op. cit., p. 2
125. Nath, R. op. cit., p. 48
126. Ibid.
127. Shahnawaz Khan, op. cit., p. 476
128. Riyaz, Ansari, p. 540
129. Nath., p. 47
130. Raunaq Ali, op. cit., p. 43
131. Riyaz Ansari, p. 541
132. Ibid., p. 544
133. Ibid., p. 546
134. Ibid.
135. Ibid., p. 547
136. Ibid., p. 548

137. Ibid., p. 549
138. Nath, R. Colour Decoration., p. 1
139. Riyaz, Ansari, p. 550
140. Nath, R., Colours Decorate, p. 69
141. Abdul Hai, p. 235
142. Bilgrami, op. cit., p. 430
143. Abdul Hai, p. 345
144. Raunaq Ali, op. cit., p. 86
145. (a) Yazdani 'Bidar'.
    (b) Consens 'Bijapur'.
146. Riyaz, Ansari, p. 581
147. Riyaz, Ansari, p. 583, 584
148. Yazdani, 'Bidar', p. 170-173
149. Riyaz, Ansari for measurement of most buildings Prof. Riyaz As. book has been referred, pg. 586
150. Riyaz, Ansari, p. 586
151. Aurangabad Gazetteer, p. 971
152. Raunaq Ali of. cit. Mss., p. 91-92
153. Ibid.
154. Ibid., p. 87, 88
155. Riyaz, Ansari, p. 592
156. Aurangabad Gazetteer.
157. Jairaj Bhog, op. cit., p. 246
158. Shrivastava Ashirbad Medieval Indian Culture, p. 142
159. Ibid., p. 148
160. Aurangabad Gazetteer Records 100 ft. Forts of India Anrid Verma records, p. 49
161. (a) Akhibate Malik Ambar 1971, p. 197
    (b) Haiq W. An Inscription in Daulatabad Fort. (Epigraphy India Maslenica 1907-8)
162. Riyaz Ansari, p. 560
163. Ibid., p. 562
164. Ibid., p. 566
165. Ibid.
166. Ibid., p. 569
167. Akhibate Malik Ambar, p. 193
168. (a) Raunaq Ali op. cit., p. 42
    (b) Aurangabad Gazetteer, p. 968
169. Riyaz Ansari, p. 611
170. Raunaq Ali, p. 48
171. Riyaz Ansari, p. 612

CHAPTER - III

THE WATER SYSTEM AT DAULATABAD FORT

"Daulatabad-Archaeological overview" is a small book on the water supply system which discusses at length the detailed analysis of the water supply schemes within the fort during the various periods or dynasties. This book written by Dr. R.S. Morwanchikar is the first analytical work based on his personal explorations that included scattered remains of fragments of earthen or stone conduits, remnants of silting cum filtering chambers, wornout portions of water towers, open channels, remains of small bunds and walls, tank-cum-wells, pressure relief towers, number of small and big cisterns and reservoirs and remains of artificial lakes at different contours.[172]

Dr. Riyaz Ansari, too is in his thesis on Daulatabad Fort has taken a brief survey of the water system and discussed the Hauz-i-Qutb, masonary work of Rang Mahal and Ab-Pash-Darra. But the entire discussion is limited in scope and does not contain the depth and systematic analytical explorations, but is mainly based on a few references. The earliest information of the water system at Daulatabad was drawn from Raunaq Ali. Another book titled "Aurangabad's Old Water Supply System of Aqueducts" gives a few tips on the water system of medieval period though of little help to the present work. Personal survey too was conducted adding to all available information collected from the secondary sources. Dr. Morwanchikar's work has been used as a role model to describe the water supply system.

It is quite evident that during the various phases of successive dynasties there was continous improvement to the existing water supply system at the fort upgrading the supply and distribution. Moreover each dynasty had to consider the increasing population in and around Daulatabad compelling them to introduce the most effective and efficient water supply.

Yadava Period—Water System :

Yadavas were the founding fathers of Devagiri fort hence the first to introduce water supply to the entire town. The fort did not mean only the fort but the nearby hills, suburbs and townships were also included within the city. Since it was a fort the ruler had to initially consider the topmost area on priority basis. As explained earlier the topmost area of the hill contained two rock cut cisterns known as Moti-taka and Hathi taka which could satisfy the needs of approximately two hundred people. It was mainly used for emergencies and not during normal conditions. Again considering the area's irregular monsoons for famine conditions sometimes there could have been periods of scarcity at the top of the hill but during monsoons it must be fairly full and adequate.

The main source of water supply during the time of Yadavas were numerous wells and huge tanks as well as rock cutcisterns with spring water or natural water which is fresh throughout the year. Whether the water was sufficient for the vast population which in all probability included Kagazipura, Shulibhanjan, Khuldabad and Nandrabad.[173] The four huge tanks which later assumed Muslim names probably had Yadava Origin due to the similarity the masonary work at Ambejogai, Kalyani, Kandhar and Dharmapur which are all famous Yadava sites.[174] The present names of these tanks are Parion-Ka-Talab, Hauz-e-Qutluq and Ab-Pash Darra.

Tuglaq period-Water System :

The Tughlaq period witnessed the transfer of his capital from Delhi to Daulatabad and it was said that the city was erected within the span of six years (1321-27 AD).[175] The city was a well planned city providing modern amenities with separate units based on occupation and profession. The fort was made stronger by adding new fortification walls, palaces, mosques and a huge township was created.

However later probably Muhammad Bin Tughlaq who had transferred his capital decided to reshift the capital to Delhi as he felt that the decision was not very wise especially due to constant invasions of Mongols. This migration had a bad effect on Daulatabad as the entire glory of the huge township bore an isolated and forlorn look.

However during this brief phase Muhammad Tughlaq had appointed Qutulaq Khan, one of his faithful brothers, to make proper arrangements for water supply for its population.[176] It is said that Qutulaq Khan diverted water from Mausala tank towards the fort and named the tank as Hauz-E-Qutuluq.

Hauz-e-Qutulaq :

A creation of Qutuluq Khan was provided with masonary steps and has a summer residence upon its bund, which now however is in ruins. Qutuluq Khan was also appointed as the Governor of Daulatabad. Though he started the construction it was completed by his brother Nizamuddin.[177] The location of the tank is near the Mausala village and is constructed in the Bala Ghat ranges. Due to the hill slopes the engineers took advantage of natural monsoon water collection by building massive dikes with steps for the storage of water. The level of the tank presently is 5 ft high which could have been much higher earlier and in monsoon season it must have overflow in the next tank of Rang Mahal area.[178] The water may be used for drinking purposes. The water was distributed upto the Rang Mahal and the Hammams etc. through strong earthern pipes. A double storey building towards the northern side, must have served for a good excursion for royal nobles and officials. Now the building is in ruins.

Bahamani Dynasty and Water Supply :

The Bahamani period offered massive architectural contributions especially the huge grand double walls of Mahakota and Kalakota, the complex of Chandminar. Some other palaces, mosques and gardens too were added.

Water supply according to Dr. Morwanchikar, [179] was provided by some wells within and outside the fort sunken deep and most of the early water system was renovated.

Water Supply during the Nizamshahi of Ahmednagar :

During this period there were numerous alterations and additions, drastic changes as Daulatabad later was elevated to the position of state capital. Naturally the increase in population forced the military commandar Malik Ambar, the brain behind all architectural alterations to revise the existing water supply schemes as well as defence strongholds. As an emergency resort he excavated a ditch around the circuit of the hill and isolated the central hill from the rest of the area. He continued excavation deep under the ground level and provided an artificial moat (60'x30') around the whole circuit.

During this period there were two sources of water supply the two reservoirs Hauz-i-Qutuluq and Darr-e-Aab-Pash. But the increased population required huge water supplies. Malik Ambar's water supply scheme was Nehr-e-Ambari.

Collection of Water through Bund :

He located the rain water collection area on the opposite northern valley as a catchment area. This rain water coming from the top was provided a small masonary diversion. Bund was constructed at the foot and along the altitude of the hill. The dimensions of the bund were 750 mm×450 mm×2000 mtrs. This served the purpose of diversion of rain water coming from the top of the mountain.[180]

The rain water thus diverted accumulated at the lower end of the diversion bund where a receiving-cum-filteration chamber was provided. The level difference between the receiving chamber and the discharging chamber (near the moat) was considerably high. Hence by means of simple gravitational force he conveyed water towards the royal residence.[181]

The receiving chamber incorporated two filtering chambers of different dimensions one below the other. The upper chamber is in brick and lime mortar while the lower one is in dressed stones. The lower chamber has two openings of different diameters where two grills were fixed. Two regular conduits one of terracotta of 20cms diameter and another of stone of 40 cms diameter were fixed at these two openings.

These conduits were laid in lime concrete with casing of fast lime plaster, brick bats and pebbles. After traversing a distance of nearly one kilometer these conduits end in a receiving tower which is at a low level from that of the above discussed filtering chamber. This tower also serves the function of discharging water and an open channel ending in the section at the foot of the escarpment runs from the opposite hill and ends into the tower adjacent to the gravitation hence these conduits work as siphons.[182] This was an hydraulic engineering feat where the laying out of the conduits and the difference of the level between these two ends were worked out without any engineering or mechanical contrivance.[183]

Rang Mahal :

Which has been mentioned earlier is a small palace with great masonary work surprisingly still in good condition. The area from Ab-Pash-Darra to Rang Mahal was also a tank. It was also built on the same patterns as Hauz-e-Qutuluq. Towards the west of the glen as the slope of the hill affords an abundant water supply of water during the monsoons where the engineers have taken advantage of the monsoons for collecting water and built massive dikes for storage of water.

Darra-e-Ab-Pash :

One of the renowed tanks is also the biggest tanks of the fort situated towards south east of the fort below the Bala ghat hills. The aqueducts built are two miles in length in brick masonary, constructed on the lowest part of the Bala ghat hills. Massive dikes are built on the same lines of Hauz-e-Qutuluq for storage of water. The tank is connected with the tanks which are situated above Rang Mahal and Hauz-e-Qutuluq. In monsoon season when water overflows it comes into Rang Mahal tank and then in the Darra-e-Ab-Pash.

The Water Supply Scheme :

This consisted of a bund to divert the run-off constructed in stone and lime followed by a receiving chamber, a well built rectangular structure plastered from both the sides with two sections for silting and filtering. Pipes were laid down underground with lime casing of fast lime concrete around both the pipes and a layer of mud. The underground pipelines and gravitational force were sufficient to carry water in the fort. Later outlets in the form of towers of 5.5 mtr height plastered from both the sides was provided to receive water from the conduits and from the top of this tower a open channel carried water into the moat, where it was stored. Lifting of water at convenient places by means of bullocks was also made. Excess water was carried to palace complex of the Rang Mahal and to the township from different direction. This was done by means of making holes in the moat wall at specific height and the excess water passed through these holes.

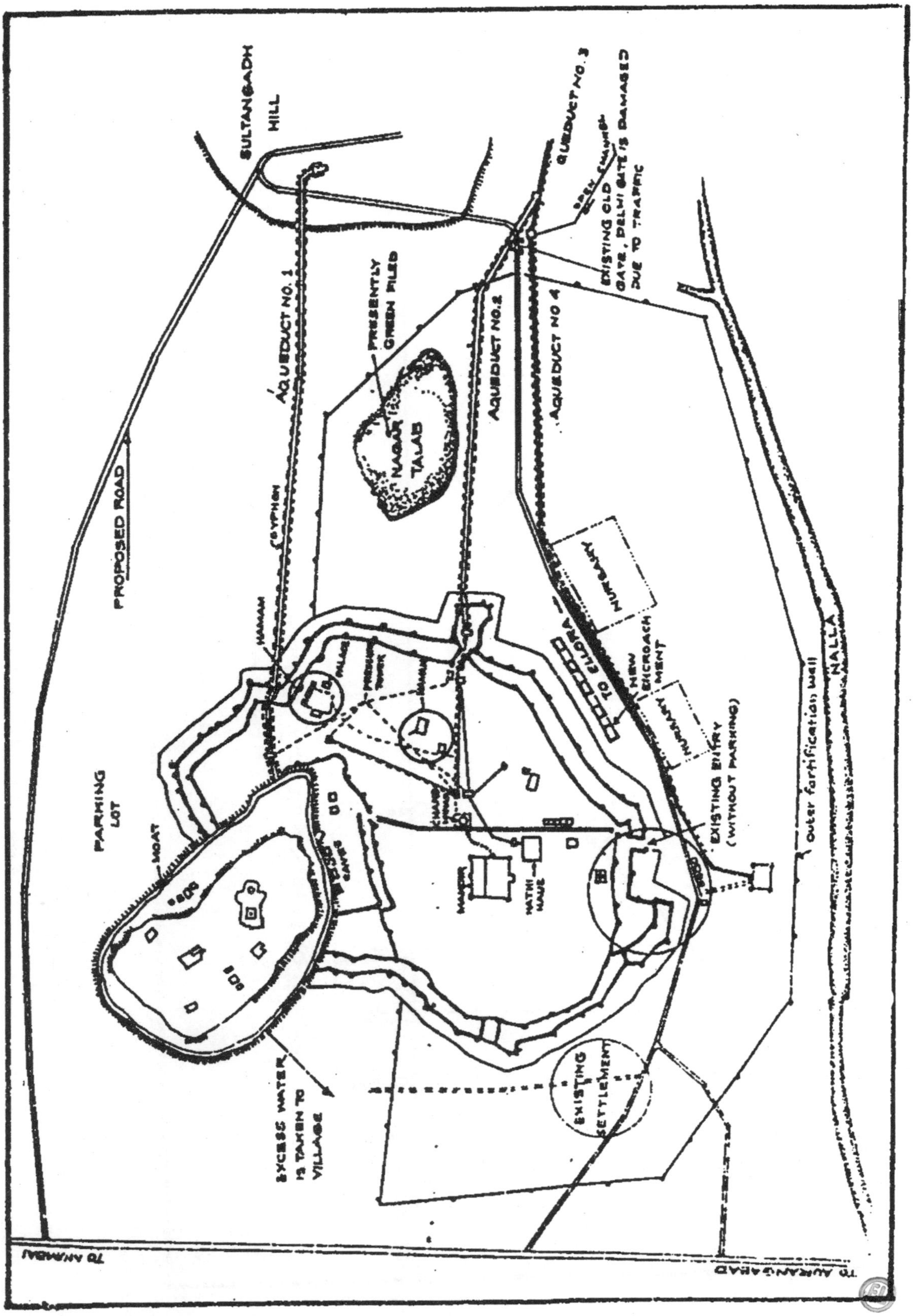

Fig. XIII. Water Supply system in medieval Period.

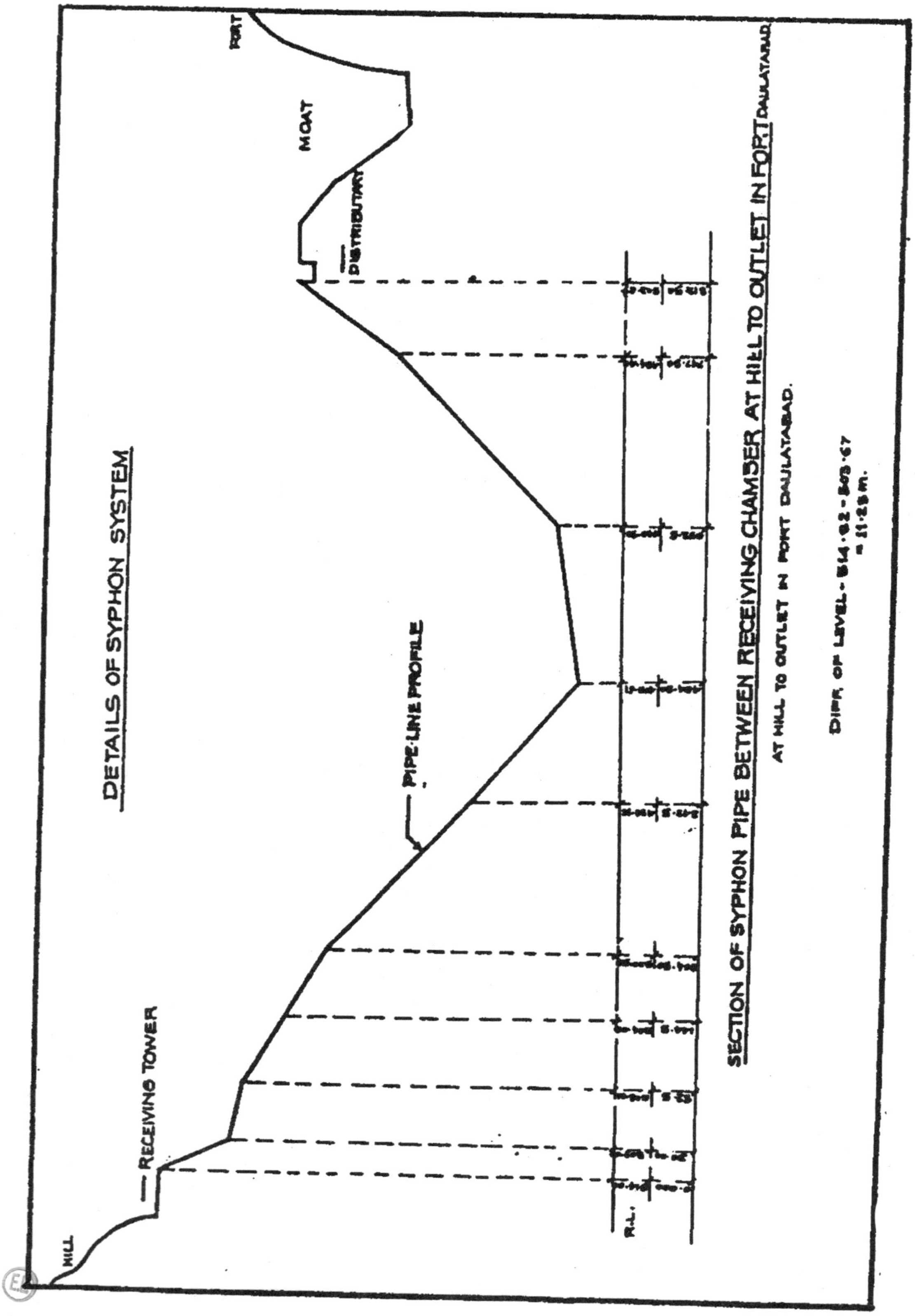

Fig. XIV. Details of Syphon system.

REFERENCES

1. Devagiri Daulatabad. Dr. S.R. Morwachikar, p. 12
2. Ibid., p. 12
3. Ibid., p. 32
4. Yahya Sirhind, 'Tarik-a-Mubarakshah', p. 98
5. Dr. Hussain, M.A.
6. Raunaq Ali, op. g. mss., p. 204
7. Riyaz Ansari., p. 601
8 Dr. Muranebha, p. 15
9. Ibid., p. 18
10. Ibid.
11. Ibid., p. 18, 19
12. Perusal Sums.

CHAPTER - IV

ENVIRONMENTAL IMPACTS OF TOURISM GROWTH IN AND AROUND DAULATABAD FORT

Interrelationship between tourism destination and environmental balance is a growing concern especially since larger number of tourists review tourists sites more seriously and consider harmonious bonds of paramount significance. Tourism as is amply evident through research studies evinces direct impact on the environment and ecology, both positive and negative. Environmental impacts are inherently and irreducibly multidimensional. If such impacts are poignant and corrosive the satisfaction levels of the visitor are pared. The modern tourist is more discriminative and not easily satiated. A subtle awareness in changing trends and fashions, comprehension of beauty, determines his judgment of a particular destination. So it is highly desirable that a tourist site meets their demand and is more attractive and appealing as it bursts into view.

Considering these close co-relations between tourism growth around a destination and its extraneous and inherent turmoil, the present study centers around a distinguished and glorified site of Daulatabad in close proximity to Aurangabad. The original name of the fort was Deogiri and a brainchild of the Yadava dynasty, once controlling a large portion of the Deccan. A hilltop fortress par excellence is considered to be an impregnable fort.

Cone shaped hill, suddenly rising from the plains to the height of about 700 ft. and surrounded by spurs of Balaghat ranges has been particularly selected for defence purposes. Its artificially excavated sides create a rocky cone, which serves as an unmistakable landmark [3] of the Deccan.

This invincible fort with an overpowering presence, a most adventurers historical fascination unfortunately is gradually succumbing under the tourist pressures. Its original natural environment is gradually modifying to human habitation which if not arrested will prove injurious and pollute the environment.211

The present study is an effort to highlight the offensive transformation of Daulatabad's present environmental hazards and pinpoint natural reserves being rapidly turned into commercially managed tourist destinations. The residents seem to lay greater emphasis on tourism development to reap economic gains whereas tourists focus on environmental protection.

Daulatabad fort since the last few decades has witnessed a dramatic increase in tourist pressure which has resulted in the degradation of its environment. Tourist increase has set profitability in the forefront and environment in the background. If this decay is not checked urgently, soon the most exceptional fort will be discredited and its importance as a significant tourist site will be ignored. The study will therefore concentrate not only on the present situation, impacts of tourism but also offer suggestions and recommendations.

INTRODUCTION

Towards the conservation and Preservation of Environmental Hazards in and around Monuments

Voluminous work has been published on the preservation of monuments, its paintings, sculpture mural paintings, palm manuscripts, but very little work has concentrated on environmental hazards in and around the monuments. Some scholars have mentioned its growing importance and the necessity of creating awareness not only amongst the concerned authorities but also the visitors and the local inhabitants whose interests are chiefly economical profits and gains. But such scholars are few in number. Again since India is a land replete with historical monuments and a profusion of historical structures are dotted in a large number of Indian states, its value is not generally recognized. Foreign countries are not blessed with such vast magnitude of heritage monuments and take extra effort to plan, manage and develop areas into perfection.

Such serious awareness does not exist in India and given a passing mention. There has been no dirth of rules and regulations, or a formulation of conservation and preservation policy either by the union or the state government. But such formulation of policies mainly are paper work for safe keeping but its application is rare. Naturally the implications of such policy formulations are not comprehended and the gravity of focusing on long term planning has not been taken up by concerned authorities.

But a new group of environmentalists mostly NGOs are breaking in on the scene and filling voids created by governmental bodies, naturally the earlier scepticism which had creeped in amongst the historians, scholars, academician and nature lovers is gradually giving way to more hope and buoyancy.

In the true sense preservation of the environment and ecology has emerged only a couple of years back. Researchers have started digging into the conservation of heritage destinations and creating an awareness amongst archaeologists, tourism departments to control adverse growth in and around the monuments especially those under the Archaeological Survey of India.

Daulatabad fort 15 km. in the north west direction of Aurangabad is one such historical site which has gradually come under the grip of uncouth development. Daulatabad's early photographs exhibit a most magnificent, imposing, formidable fort rising to heights from the flat plains, enriched with a variety of green vegetation in chaste and unaffected environments. As it burst into view the early native charm of the fort has transformed into vulgar structures, and been deprived of its virginity and purity. The ugly tin shacks called restaurants, dowdy electrical poles, trucks, rikshaws create an unkempt environment.

As we turn right towards the main gate right opposite the fort, another set of structural growth, an unsightly tangled mass of eating outlets dot across, creating more disorder. The Indian visitor has accepted this growth as natural as the sight is common around all national tourist destinations. The foreigner who is used to neat, polished, embellished tourist sights are shocked. For them it is a cultural shock. The difference is so deep, the contrasts so extreme and its purely out of our attitudes of 'Chalta Hai.'

Methodology :

The approach for the major part of the study is based on field surveys. Personal observation aided by photographic study has proved of immense help. Several trips to the fort, in and around the fort has revealed a number of problem areas. Problems which require urgent attention. The huge area of the fort within, without, the periphery, was included. Roving and rambling around the fort and a detailed structural study of the number of palaces, temples, masjids and their present condition was investigated for this study. The studies have examined and fathomed the change in physical appearance of the original environment into a creation of disturbed environment wherein the restaurants clash with the beauty of the fort.

A sample survey of officials from Archaeological Survey of India, Western Circle, Govt. of India, Department of Tourism, Maharashtra Tourism Development Corporation, members of History Academy and a few environmentalists were interviewed. The samples totaled twenty five from the Government agencies and twenty five NGOs. The results were interesting, as the government officials were more concerned with impressing their positive contribution, while any negative functioning was simply evaded. The NGOs concern at Daulatabad, which they insisted should be dealt with on emergency basis.

The project has also included a few secondary sources. Books are limited on the subject but a few books recently published have proved helpful. Krishan Kumara's book 'Managing Tourist Destination' does highlight certain environmental issues, like his analysis of Social and Environmental Patterns discusses tourisms relationship with ecology and environment which he confesses are the primary features contributing to the attractiveness of a tourism region, and it has both negative and positive impacts on the destination. Another very recent publication of

Jagmohan Negi, 'Foundation for Tourism Development' also discusses the attitude of the Host Population at tourist destinations. Though he has dealt with the environmental problem to a very limited extent his suggestions are helpful.

Two articles on conservation have dealt with conservation policies of Government towards national monuments and an article in, Lokmat Times on "Conservations as a Cruel Joke" have offered important guidelines for this study. The author Rupika Chawla's, 'Conservation as a Cruel Joke' centres around the negative aspects of conservation at our heritage rights especially under the A.S.I. Another book by A.K. Seshadri 'Conservation of Monuments in India' has though concentrated mostly on conservation of paintings, sculpture etc. also dealt on the Preservation of Environment and Ecology around the destinations though it is a passing reference.

Findings :

After going through primarily in the early stages through some A.S.I. rules and regulations which examined the permissible distance allowed between an A.S.I. recognized tourist site and eating outlets, and also the general distance that is advised from the monuments, there were a number of revealing factors that came to light. As I have mentioned earlier my findings are not only based on the A.S.I. rules and regulations, and samples from A.S.I. officials and other Government officials an on the spot exploration, investigation and corroboration of facts from NGOs was used for a balanced and analytical measurement.

There are several rules and regulations to protect the monuments, its sculptures and paintings. Though a few of these regulations lately have also included areas around the monument to be conserved as protected area. The central government brought an amendment to the act which was made public on 5th July 1992 by declaring areas upto 100 metres from protected limits as 'prohibited area' and further beyond it upto 200 metres as 'regulated area' for purpose of mining and construction.' John Marshall's conservator ethics initiated that any innovations should be made as per the needs of the time without damaging the original personality of the monument[2].

In conservation of monuments a holistic approach is necessary as otherwise the concentration would be directed only at conservation of the actual structure while the environment, if uncared for, would give rise to unbalanced violent juxtaposition of unaesthetic architecture. One cannot ignore the inter-connections and isolate environment from the general conservation and preservation norms.

Presently the Archaeological Survey of India is the sole controlling body in charge of the entire conservation, management process of a large number of monuments spread all over India. Considering the magnitude of monuments under A.S.I. in each circle it would be near impossible to manage, plan, develop and execute the plans. Radical restructuring is therefore absolutely necessary to meet the increasing demands of conservation.

Now though rules and regulations do limit activities or rather building and parking activities around a monument upto an area of 100 to 200 metres as prohibited and regulated area, it has not been specified whether the law is applicable to all A.S.I. monuments or only a limited number of monuments. As the act is infructuous as infringement is often the rule due to unawareness of the act. This rule has rarely been implemented. The continuing growth of population and illegal spread of settlements have made the task of preserving monument a difficult task.

Daulatabad Environmental Impacts :

What is reshaping the periphery of Daulatabad fort is threatening the very landscape of the fort. In the future this growth will be disastrous and the outcome catastrophic. As a visitor first views the fort from close quarters his complete view is obstructed by numerous shops, vehicles, restaurants, high rising electrical poles which obstruct a complete view and the satisfaction level is reduced. Quality enjoyment at viewing a heritage monument is natural surrounding which is lacking.

Some of the structures are so close to the fortification walls that the periphery wall from the left side of the fort is hardly visible. Small cement buildings are only a few feet away from the fort. Naturally the norms of rules of prohibited or regulated area around the monument hardly exists at all.

Again with the growth of such structures a gradual degradation of the environment is but natural. Huge disposal of garbage and litter is evident right in front of these shops, transforming the entire landscape into an eye-sore.

Besides, as the visitor goes further down to the right, he again comes upon a host of eating outlets on the opposite side of the fort, fringing the entire road. There are several such outlets, each competing with the other for attention from the tourists. With eating outlets naturally litter is a co-related factor. Both go hand in hand. The condition is still not as traumatic as other metropolitan cities. But in case there are no controls the situations can get out of hand.

Secondly the adjacent area, immediately outside the walls of the fort on either side of the main gate, is exploited as parking space. It is an authoritative parking space. Besides lighter vehicles like scooters, motorcycles, rickshaws and cars, heavy vehicles like buses, trucks and tractors also utilize the area for parking. The distance is just a few metres from the fortification walls. This sort of proximity can create perpetual vibration, which has been continuing since decades. Vibrations are responsible for displacing the fortification stone bindings. Besides the continous emission of dust, fuel also corrodes the lime binding. In fact due to its strength the lime has clasped to the bricks and stones since centuries but modern day vehicular movement is threatening its very existence.

A third problem area, which has specifically vexed the NGOs, are the brick kilns exactly behind the fort, bordering the Chand Minar. The regular row of bricks mounting higher and higher and the smoke emitting from the kiln is clearly visible. The Chand Minar appears to be nearest to the kilns. The smoke keeps settling on the fortification walls and the problem is harrowing all serious scholars, historians and NGOs who are genuinely concerned (Plate XL A and B).

The fourth problem are the number of structures spread around the fort within a radius of two kilometres which consist of building from Muhammad Tughluq's period. Muhammad Tughluq who had transferred his capital from Delhi to Daulatabad had created an entire township. The township is mostly in ruins but some of the beautiful structures which exist even to this day are illegally occupied, without any maintenance resulting in cracking on the structures and vegetation growth above and around the domes. The beautiful fabrications if not immediately protected might collapse one fine day.

Fifth problem area is the amount of sand being removed from all the area around the fort for brick kilns. Truck loads of sand is removed. Daulatabad in fact is famous for good quality bricks and people from Aurangabad buy in huge quantity and bulk orders are placed with the owners. The consequences are that due to increasing demand more and more brick kilns are mushhrooming. This is resulting in rampant removal of sand causing irregular, uneven surface covering which is clear from the photograph especially clearer as the area around the pole makes it evident as to the amount of sand that was removed for brick kilns. This is leaving wide scars all around the fort area.

A very urgent issue is the tall, soaring height of vehicles carrying scooters, motorcycles and cars passing through the Ambarkot or the last fortification wall close to the ghat. The height of these vehicles is so high that the medieval gate, which was created for medieval period transportation, does not agree and harmonize with the modern requirements. Very recently similar high-rise vehicle carrying scooters rammed with the gate and just couldn't come out. There was a traffic jam and for hours together the traffic could not budge. The result was destruction of this renowned gate partially. As the entire area around the city of Aurangabad, the towns of Daulatabad and Khuldabad are strewn with such celebrated gates, tall vehicular movement should either be prohibited or some alternative arrangement should be made for regular movement, which will not prove hazardous to the medieval monuments.

Interior Problem Areas (within the fort) :

Besides Daulatabad's periphery there are several problems within the fort which are tormenting the NGOs. One of the problems which probably is not really noticed by either the A.S.I. or any other related government bodies are the number of beggars lining the entry area of the fort creating a negative image of the destination. Added to the tourists woes are the number of touts and hawkers selling cheap literature. There is an increasing competition amongst these human groups to harass, the visitors, run after them, pull at their sleeves, beg of them to extract as much money as possible from the tourists.

The first fortification wall through which a tourist passes has collapsed partially and its gates have disappeared. But the fixture openings are still visible on either side. A wooden piece is evident which is gradually gnawed by the termite. The opening is also being used by the average tourist unaware of preservation for dumping litter. Chocolate covers, empty plastic bottles are callously discarded creating a very bad impression.

On the entrance through main gateways as one enters the huge spacious courtyard, on the right hand corner is another closed gate. Within the gate the constricted area is utilized for garbage dumping, parking space for the employees, and storage space. Huge thick ropes from medieval period and a medieval object (either nagara or a cooking utensil) is stored in the arched openings. The huge, thick rope is unique of its kind and definitely used during the royal residency period. This rope can be exhibited as a piece of museum object instead of allowing it to rot in this forlorn corner of the fort.

As one gets through the several labyrinthine, circuitous fortification walls, one is astounded at the formidable, imposing walls which rise to perplexing heights. The walls also display beautiful ornate pillars, oriel windows, aesthetic stylistic animal motifs and floral motifs. The innumerable fortification walls are in a sorry state of preservation, as several stone boulders are loosing its grip and could land on anybody's head for that matter. Very little attention seems to have been paid to the top level structures.

Palatial Building (palaces, hammams, masjids, temples etc.) :

Palatial building and their problems of conservation since Daulatabad has been under the domination of at least seven different dynasties it is but natural that each one of them leaves memorable imprints of their royal period, each distinctive in its own architectural styles. The earlier chapters have dealt in detail with archiectural remains of Daulatabad.

Hammam :

As the visitor moves towards the Chand Minar, in the interiors on the right side, (back side of Chand Minar) is a beautiful Hammam in a bad state of preservation. On the domed ceiling of the hammam outlets are provided for probably for ventilation. On the ceiling several threadlike cracks have appeared which with time will gradually get wider and wider resulting in the collapse of the hammam. Though the hammam has been well kept and is clean, maintenance does not simply include cleanliness but also restoration, and timely restoration.

Fig. XV. Various lakes reservoirs.

Fig. XVI. Pavilions (Diwan-e-Aam).

Diwan-e-Aam :

Further down the same pathway the visitor comes across a beautiful medieval structure popularly called the Diwan-e-Aam. Recently the palace has been whitewashed. The most humorous part of this white-wash is that only the portion which is on immediate public view is white washed while the sides are left to the dogs. Besides the color application, even new doors are fixed giving it an altogether modern look. Beautification is often used when walls are whitewashed[3] and the cracks and other drawbacks are shrewdly concealed. Generally the more discerning will probably be astounded with the lime color and the modern doors and in short completely overlooking the exact sense or meaning of restoration. Unless any kind of repair work especially heritage monuments are not well researched, its original layout, intention, color schemes have to be studied by an expert before taking up any restorative work. For any kind of preservation the environment, the features such as size, scale mass, color and materials of buildings should be preserved.[4] In India monuments under A.S.I. are treated with scant care, or careful research work to support any alteration in it's addition, repair or color schemes and the result does not in any way agree with it's medieval surroundings. The palace appalls the professionals and to him would manifest a complete vista in discord with its environment. If a color scheme which went or harmonized with its environment was applied it definitely would not have caused so much of anguish to the honest conserver.

Masjid :

Still further down is a small but graceful masjid supported by decorative pillars and three arches. The upper portion has partially crumbled down. Grass and trees are not only growing around it but are crowning it. The interiors of the masjid are in a bad shape and need urgent repairs (Plate XLI).

Chini Mahal :

Emerging out of this portion of the fort, the visitor ascends the main gateway staircase and arrives at the right side and comes before another palace which is popularly known in the vicinity as Chini Mahal or the place where once Tana Shah the last ruler of the Qutb Shahi dynasty was imprisoned by Aurangzeb. The palace was far larger than what remains. What can be viewed to-day is only the skeleton of a once grand palace. It is called as Chini Mahal due to glazed tiles (blue and white) which once decorated the palace (Plate XLII). Faint remains of the glazed tiles are visible even to-day. From its sideways cracks can be seen clearly and bricks and stones are on the verge of crumbling any moment. Similarly at the lower level the pillars have formed cracks and these cracks are growing wider. The mass of stone and brick bindings have already splintered and crumbled to dust. If the remains are not preserved the whole place will soon perish leaving no memorial for the future.

Rang Mahal and Sonheri Mahal Yadava Darbar :

The (Plate XLIII) names of these palaces have been (Plate XLIV) passed on and its authenticity is doubtful. These are two beautiful palaces in the worst state of preservation. A mere glance around these palaces offers vistas of pure royalty, with graceful arches, engrailed arches, arcades, floral motifs, vines, pavilion decor, refined deep mehrabs, woodwork, wooden pillars enriched with finery. These exquisite elegance is languishing. Thick growth of grass projects from every nook and corner, from its roof, walls, mehrabs causing a disenchantment with the conservation work of A.S.I. Examining this crumbling royal palace one would wonder what exactly are A.S.I. concerns ?

Even the wooden rafters which even a year back were in a fairly good condition, today are in a distressing condition. Wooden beams have broken loose from the ceiling and are swinging merrily awaiting the destined victim. A couple of years back these palaces, slightly isolated were used for toilet purposes as no toilets are provided anywhere near or inside the monument. During this visit the interior looked cleaner. Maintenance for the A.S.I. seems to be to keep the surroundings clean, but allow it to crumble.

Graffiti work :

The public at large simply loves to inscribe their names, the names of their beloved on the fortification walls, inside and outside palaces can be seen rampantly. Graffiti work is at its height on Ajanta paintings in cave 10, where several early visitors had a heyday inscribing their wonderful names. Though at several places the graffiti has been meticulously rubbed off by the A.S.I. but their careful work seems to have no effect on the enthusiastic visitor who leaves his name for posterity just like people have built monuments.

As one ascends the staircase of the Mendha topp a beautiful vista of an enchanting landscape spreads before you. As the eyes rove from left to right they suddenly settle on a resort (Deogiri resort) which has recently cropped up. The juxtaposition, the projection is so acute that it simply irritates the environment.

In fact it was against such projections that Maharashtra State passed laws. The state government has prepared a Revised Draft Building By laws and Development Control Rules for Greater Bombay and sanctioned the Development Control Regulation No. 67 alongwith Appendix VII A from 1st June 1955. The main provision in the notification under para 7 relates to maintaining skyline. Buildings included in listed heritage precints shall maintain the skyline in the precinct as may be existing in the surrounding areas.[6]

Discussions and Suggestions

There are umpteen such problems in and around the fort. Since tourism has a very strong, intricate but fragile relationship with ecology and environment the primary features contributing to the attractiveness of a tourism region should be protected. The quality of the destination, its purity and natural charm will determine not only the type of tourist, but also tourist traffic flow patterns and effective and potential tourist activities, but also the overall magnetism of the given destination.

Though very recently few attempts are being made to beautify the surroundings of the fort the efforts have been mostly directed within the interiors of the fort. The eating outlets which create an unhealthy effect and extract the very life breath of this beautiful fort can be transformed into an aesthetic enchantment. The trouble is with the growing population of Aurangabad which encourages the spread of illegal settlement in its periphery and the task of preserving a cultural heritage becomes difficult. The main pressure is from slums, builders and muscle power which has contributed to the careless growth of ugly structures around the fort. What is required is the immediate removal of such illegal structures. This can be achieved through certain urgent measures.

1. The local residents who own the shops should be taken into confidence.
2. The participation of the local residents should be mandatory.
3. Alternative arrangement for housing shops and restaurants should be made.
4. Professional bodies, NGOs should be involved in the process.
5. Architects with some background in restoration work or architects with research work on concerned subjects should be involved.
6. The shops and restaurants should be built with architecture which blends with the environment and color schemes, height and material should be in keeping with the landscape.
7. Material used for these structures should mostly be local.
8. Planning, development of a model site should be studied and researched to avoid blunders.
9. No building should be erected to spoil the existing character of the environment of the monument.

In fact the preservation of all tourist monuments poses immense problems because the public in general is still not aware about the norms of conservation, and the academicians, archaeologists, politicians and local residents hold varied views. Some want that all the monuments should be whitewashed, some want that what is missing should be replaced and others prefer structural stability and beyond that nothing. But a planned process of conservation could be adopted if tourism is to succeed. A number of points should be taken into consideration.

a) The ambience of the landscape

b) Tranquillity of the atmosphere.

c) Parking at a safe distance

d) Alternative road and traffic movement.

e) Scientific garbage disposal methods.

f) Infrastructural services of a high standard.

g) Pure water and toilet arrangement with proper signage.

h) Nature of buildings.

i) Interior setting and ambience to conform with the environment.

j) Exterior layout to add to the beauty.

k) Planning controls.

l) Continous monitoring.

m) Future building regulations.

n) Public awareness programmes

o) Annual repairs and maintenance.

For this kind of processes a scientific study of the monument is essential which would primarily include complete documentation of all the present fortification walls, palaces, temples, hammam, masjid, moat and all the noteworthy features of the fort like moat, the spiral subterranean passages, the water system, all the secret hidden passages from where the enemy was attacked.

Future Plans of A.S.I. :

The Archaeological Survey of India too has future plans to develop and beautify the monument, but none of these plans include the exteriors. Their plans include creation and designing of lawns and gardens inside the fort. In fact the A.S.I. has done a commendable job inside and developed several lawns and gardens which are also regularly maintained. Their plans also include a museum, wherein the Diwan-e-Aam which has recently been whitewashed is planned for a museum. There are several artifacts, remains of temple which will be displayed aesthetically inside the museum. They also plan to attend to structural conservation wherever needed.

Where is structural conservation needed is a million dollar question. Do these trained A.S.I. officials need a layman's advise for where conservation is urgently required? Are not the marks of decay amply visible? Conservation is not only an important function of the A.S.I. but it is an urgent demand. In case brick by brick of all the structures collapse the very existence of A.S.I. will be under threat.

One must remember that India's identity is not only a single Taj Mahal. There are several other monuments of equal stature which too require equal and serious attention of all top authorities. If our structures of heritage value start disappearing our very identity as a heritage country will come to an end. The general excuse given by A.S.I. officials are always lack of funds. Now the rate of entry tickets for foreigners and Indians has increased multifold. Again India can also involve global partners like UNESCO for the preservation of a few unique monuments like Daulatabad. Other interested organizations can also be invited to offer expertise and create an attractive environment for a gratifying experience.

REFERENCES

1. Dikshit K.N. Towards the formulation of national policy on conservation of monuments, p. 786
2. Ibid., p. 788
3. Chawla Rupika, 'Restoration needs Restoration' Lokmat Times. January 10,2000
4. Seshadri, A.K. Conservation of Monument In India Book India Publishing Co., Delhi 1991, p. 150
5. Dikshit, K.N., p. 788
6. Ibid., p.788
7. Kamra Krishan, Managing Tourist Destination, p. 25

CHAPTER - V

MONUMENTS AROUND DAULATABAD FORT

Ibn-e-Battuta the chronicler, traveller and Chief Qazi of Muhammad Tughlaq while travelling from Ujjani to Daulatabad delineated it as an enormous city which rivals Delhi, the capital in importance and in the spaciousness of its planning.[1] Thevenot described it as indifferently big and is much longer than broader.[2] Daulatabad was an important part of the riyasat of Deccan as it was the capital during Muhammad Tughlaq's period and later most of the succeeding dynasties used it as an important military base. It's limits were vast upto Ellora from the north as all the villages in between were included within its limits[3]. The different dynasties from Yadavas in 1187 A.D. who were responsible for the excavation of the strong fort, the Khiljis, the Tughlaqs, Bahamanis, Nizam Shahi dynasty of Ahmednagar, the Mughals, Marathas and the Bizam Asaf Jah-I of Hyderabad left their imprints in the form of structures big and small. The earliest dynasty which founded the fort had concentrated mainly on the fort and its defences. But the first Muslim dynasty which invaded the fort was Allauddin Khilji who had camped for several months around the fort and during this period and later during his several invasions by Malik Kafur and finally the complete annihilation of the Yadavas the Khiljis have built a few buildings around. Muhamad Tughlaq probably left more remains as he had trensferred his capital from Delhi to Daultabad. Their successors the Bahamanis were great builders and their contribution in architecture is well recognized.

The monuments around Daulatabad were both religous and secular as in the interiors of the fort. Those of the religious nature of building were of two kinds mosque, Eidgaah, and the tomb. While the secular buildings consisted of palaces, caravansarais, shikargah hammams, water tanks, water system, shahi musafir khana, Hathi Khana etc.

The religious structures were the mosque or masjid, literally "the place of prostration: it is an important building of faith and also the key note of the style". Originally the masjid started as a simple architectural building but gradually it developed into a most complex and intricate style. The first structure which developed is today called an Eidgaah generally used only for Id prayers as large number of devotees congregate in one place and the space required is huge. Again Eidgaah was the early masjid when prayers were performed in the open air with nothing between the devotee and his God's[5]. However regular prayers had to be performed in more protective area due to natural climatic conditions. The actual masjid consists of a rectangular open space or (Sahn, Courtyard) the four sides having pillared cloisters or liwans with a fountain or tank in the centre for ablutions, a ceremony described as "the half of faith and the key of prayer". To meet the demand for some focal point in the scheme, the cloisters on the Mecca side (in India on the west) of the courtyard were expanded and elaborated into a pillared hall or Sanctuary, with a wall at the back containing a recess or alcove called a mihrab indicating the qibla or direction for prayer. On the right side of the mihrab stands the mimbar or pulpit while a portion of the sanctuary is screened off into a compartment for women. An elevated platform from which the Muezzin summons the faithful to prayer is also a necessity, and usually takes the form of a high tower or minaret. Besides several mosques in the interiors of the fort one comes across several masjids in good condition around Daulatabad fort.

Eidgaah :

Of Alauddin Khilji's period. The general concept of an Eidgaah is to accomodate large number of worshippers during both Ramzan and Bakri Idd. The Idd namaz main intention is to allow space for thousands to perform namaz at one time as well as to meet and wish each other at the same place after namaz. Since a regular building is limited in space a plain wall with a pulpit for the Muzzin is built. The wall generally consists of a recess or an alcove called a mihrab indicating the quibla or direction of prayer. On the right side of the mihrab stands the mimbar or pulpit. An elevated platform from which the Muzzin summons the faithful to prayer is an important part of the Eidagh[6]. The part where the mihrab or symbol of direction of the faith was the most significant portion of the Eidgaah.

One of the most beautiful Eidgaah in Maliwada behind the fort of Daulatabad was built probably during the period of Alauddin Khilji or the later Khiljis as is evident by the style of architecture. The inventiveness of the arches, the method of its walling is very similar to the Alai Darwaza in Delhi (1305) built by Alauddin Khilji. Although there is little correspondence in the actual form, yet the example of same general texture, aims at the same effects and is animated by the same spirit[7] as the Khilji art. What makes the architectural style so special at the Edgaah are the type of arches very peculiar to the Khilji style known as pointed horse shoe or 'keel' or rare kind of arch not ordinarily used later as this particular kind of arch unfortunately does not continue after the Khilji period hence the conjecture that the Khilji style continued later would be absurd. It has been lent a similar decorative treatment like the Alai Darwaza though it lacks the marble decoration. There are two outlines the upper with a pointed horse shoe arch while on the underside or intrados is a fringe of spear head. The arch is contained within a rectangular framework which is plain and not decorative like the Alai Darwaza (Plate XLV).

The wall of the Eidgaah is decorated with five arches with the central arch slightly broader and higher than the side arches. In between the central arch and the side arch towards the right is the staircase for the Muzzin. The Imam stands here and offers Qutba (Sermon) after which he descends and stands within the central arch. The member and mihrab are a regular feature of the Eidgaah. Here the Idgaah also has an added feature of a small room above the wall with three arched openings while the back side is closed. The room is crawned with a small rounded dome. On both the sides of the Eidgaah are rounded bases crowned with towers on either side. On it's backside is a staircase to reach the room above. There is a wall—Walk on the wall where a single person can easily walk. Four steps are constructed on either side of the wall above to reach the rounded domes above. The Eidgah once had a enclosing wall. The enclosing wall is still very much visible but parts of this wall have collapsed. The central entrance gateway have square pillars also crawned with small domes which lends the gate special beauty. The surrounding wall is very low in height (Plate XLVI).

Jama Masjid :

At the backside of Daulatabad a little distance away is a Jama Masjid in complete ruins. During its heyday the masjid must have one of the most beautiful one. What remains of this masjid are walls on all three sides while the front side has completely fallen, though the lower portion is still visible. The roof has fallen as nothing remains of the roof. Since the structure is within a private farmland from a distance it is hardly visible, more so since it is camouflaged within thick trees. Only when one comes very close is the structure visible partially.

The back wall of the Jama Masjid is decorated with five slender blind arches very similar to the Khilji arches. There is one opinion which places the masjid to the Tughlaq period. But the arches of Tughlaq period are four centric while the arches of Khilji period have a pronounced horse shoe shape arch. The wall consists of blind arches with the pulpit in the centre.

On the right side wall are open arches with steps to ascend while the left wall also has slender open arches. Remains of three columns once intricately decorated are still visible. The masjid is stretched lengthwise while it is limited in breadth.

Once the entire masjid was supported on columns of which only remains are visible. There is no attempt at restoration as the masjid is under private ownership. The masjid is completely built in stone and is the finest example of the 13th century A.D. (Plate XLVII).

Khufia Masjid or Secret Masjid :

A very unique masjid inside a well. A novel concept wherein the central portion of the deep well is a masjid built in stone. The entire portion above the wall has been hollowed or excavated on all four sides and an entire masjid is built with space distributed equally on all four sides (Plate XLVIII).

From a distance nobody would even suspect that a monument of this type exists. As one comes closer to the well steps to descend inside the well are visible which lead to a masjid as you turn to the right or left. The masjid is excavated a little above the well in the centre and has three arched openings on all four sides. It is in the square shape.

In the interior on all four sides are blind arches with the pulpit in the central portion for the Muzzin and below are rive blind arches in a semi-circular form.

There are pilasters supporting the narrow roof on all four sides. The entire masjid seems to have been recently repaired as is evident from the plaster work. Above is the chhajja (ax projecting stone feature to protect it from rain and sun) with plain rafters. Below in the centre is a well with water available the year round.

The Khufia masjid is a unique concept not seen anywhere and there are hardly any example of this type of masjid. Its period is difficult to understand. Though the arches are simple and seem to be similar to the Tughlaq period it would be too presumptuous to definitely ascertain its date.

Qutlaq Khan's Shikargah :

Towards the north of the fort on the way to Kagzipura is a rough pathway leading to a beautiful structure popularly known as Qutlaq Khan's Shikargah. Qutlaq Khan was the governor of Daultabad appointed by Muhammad Tughlaq. Qutlaq Khan's contribution in public utility work and social work received tremendous responee from the people. He was so loved by his riyasat (subjects) that when Muhammad Tughlaq recalled him to Delhi the people were so disheartened that they saw themselves on the brink of ruins. They had enjoyed a period of tranquillity under the just and benevolent rule of the nobleman.[8]

The Shikargah attracts immediate attention and is visible from a distance. The architectural features of the Shikargah are prominently that of Muhammed Tughlaq period. The building is a stone construction, the walls are slightly battered and the arches are no more the pointed arches of the Khiljis' period.

The most significant and eye-catching feature of the structure is the projecting oriel window or hanging balcony which unlike the later period is not hanging but built at the ground level in a projecting form. The chhajja above this projection is supported by brackets four in number and two brackets on the sides. The window has three open arches with full blown lotus medallions on either sides of the open arches (Plate XLIX). From the interior this portion of the window is on a slightly elevated position. As one enters from the front side on the banks of the tank there is an arched entrance to the main hall. The long hall is supported on columns with two arches. In the interior portion are square mihrabs. The arches here are no more horse shoe shaped but have been modified into tudor outlines and there is a slight ogee curve at the crown.[9]

To first floor unlike the lower portion is built in bricks. There is a staircase from the front portion towards the left side. As one ascends one comes across a long hall which has a side room, with a vaulted ceiling. There are several big and small mihrabs within the wall. The room on the right side appears to be an addition during the Nizam's period and iron beams are visible on all three sides. There is another hall opening at the back with five tudor open arches. The roof has collapsed. Most of the building has open arches which served the purpose of hunting. As the shikargah is on an elevated position the country around can be viewed very clearly. (Plate L).

It was an apt place for bunting and there were countless flocks of heron and other game targets. Aurangzeb and his sons Mauzzam and Azam Shah delighted in hunting nilgai and the heron. The nilgai were shot from the fixed status of shikargah as they were driven down the narrow valley and the herons were struck down by trained hawks.[10]

Rang Mahal or Shahi Musafir Khana :

Is a construction in complete ruins exactly behind Daulatabad fort and is said to have housed the royal guests or the nobility. The huge area around the Musafir Khana was enclosed by a high wall most of which has collapsed but enough remains to give an idea of the height of the wall which was sufficiently high. The main building is in complete ruins. Within the fallen wells and roof a few arches are visible. It was a two floored building. The lower floor has small windows while the first floor opens on either side by arches. The Mahal within the interior portion was divided into rooms and the backside wall was decorated with blind arches. The building is located on a small hillock and is built on a raised platform. Due to its elevated position it is visible from a distance.

In close proximity to the Rang Mahal is the Hathi Khana or a structure to house elephants. This is also a magnificent building in rectangular shape with blind arches on its interior and exterior wall. The span in between the arches are wide and hence its closer affinity with the Tughlaq period.

The interior of the Hathi Khana is in a fair state of preservation and consists of two side rooms both with an open arch and the right with plane arch pointed above. The central portion is vaulted and is decorated with the most attractive designs in diamond shape. There are mihrabs on all the three sides of each hall and the central portion too is decorated with mihrabs. On either side of the entrance gate are steps to ascend. Above in the centre is a dome. The plaster in the interior portion is in a fair state of presentation (Plate LI)

There a were tanks on either side of the main building. There was an underground water channels which provided water to the guest house. The brick construction under-ground with pipes broken at several places are visible even today.

MAQBARAS

1. Maqbara of Shia Saint :

On the way to Daulatabad from Aurangabad is an imposing maqbara of Shia Saint. The maqbara is not visible from the main road. One has to walk for sometime and after a five minute walk there is a river which is crossed to reach the maqbra. From here the maqbara is visible which appears solemn and majestic especially due to its location of hills in the background and lush green farm land spread behind. Within a low enclosure a building complex can be seen. The enclosure consists of low walls with four burj or towers on all four sides. Except for one the rest of the towers are in a bad condition but enough exists to understand the refinement and elegance of the burg. The burg were supported on rounded tall bases. The topmost portion consists of a battlement design. The main gateway has a plain arch.

The first building is built on a lower plinth and is a walled room. It has three open arches on either side while the closed walls have blind arches. There are mihrabs in the wall. The ceiling is partially broken.

On a higher elevation is an open courtyard with a khankha on the left and the maqbara on the right. The khankha has five open arches. The arches are plain with a broken chhajja on the top supported by plain brackets. The arches are depressed four centred arches and is reinforced with a supporting beam. This arch beam combination is a hallmark of the Tughlaq style. It has the ability to span wider spaces.[11] It is typical of the masjid which specially provided for the traveller paying worship to the religous leaders dargah.

Opposite to the khankha is the maqbara of a Shia Saint. It is an imposing building walled on all four sides with an opening towards the north. It is an impressive domed structure. From the exterior it is carved with graceful decorative double pilasters on all four sides of the buildings. The pilaster has a square base, two steps with a rounded member above and an octagonal shaft. The capital is Amalaka with a number of members above. The pilasters are placed one on top of the other with similar design. It appears to be double storied building with three engrailed arches on the lower and upper portion. Within the lower arches are small hallow mihrabs. The upper three arches are plain within. There is chhajja surrounding from all four sides which are partially broken. The building is crowed with a huge dome with a neck, superimposed by an inverted lotus and a kalasa. The entire maqbara is made in stone (Plate LII).

From the interior of the tomb, to support the vaulted dome are squinches that consist of projecting arches. This is a contrivance across the upper portion of the angles of this square maqbara. They are placed diagonally at the internal angles of the square room in the phase of transition to convert it from the square to octagon to support the circular dome.[12] In the centre of the square maqbara is the tomb of the Shia Saint.

Towards the south on a lower side is a well called Tharola a typical example of the Mughal wells. It is plastered within. The water comes through earthern pipes and even to this day there is ample water. There are steps to descend inside the well and near the lowest well the entry is through an arched door.

Probably the entire building complex belongs to the Mughal period as is evident from the dome architecture the engrailed arches and the developed squinch supports.

Tughlaq period Maqbaras :

In close proximity to the fort the main road towards the left two magnificent maqbaras can be witnessed. In between the two maqbaras is a notice board from the Nizam's period calling attention to these two tombs as protected monuments. It is written:

"Protected Monument under the provision of the ancient monument Act V I H VIII 1337 Fasli. Any person who damages or disfigures this monument is liable to imprisonment which may extend to three months or a fine which may extend to Rs.500/- or both."

Ghulam Yazdani,
Director of Archaeology.

The First Maqbara :

Appears more decorative. The structure is built in bricks and once was completely plastered. Though even today a large portion is plastered. It is square in shape with typical Tughlaq type tudor arches, four centric. The arches on all four sides are open and recessed. Another feature of the Tughlaq period was the element of reduced ornament, confined mostly to inscribed borders and medallions in spandrels executed in plaster. The decorative features of Khilji period like geometry and arabesque are hardly evident in the Tughlaq buildings. As in the Tughlaq buildings there was a distinct charge in the spirit and tone of architecture. The change was from the elaborate ornamentation and luxurious splendour to one of Islamic simplicity and austerity. The desire to avoid ornamentation and richness in details might have been probably due to economical reasons rather than religious puritanism[14] (Plate LIII).

The brackets of the rafters above are still visible. Above the Maqbara are four burj in the same shape as the main structure with a rounded dome. The central dome above is rounded with slight inverted neck. The interior of the maqbara is vaulated. The lower portion consists of floral decoration in its corner. There are squinch arches on all four corners. The stalactite (drippling stone) design is visible. It is a type of arnament resembling the formation of calcium carbonate hanging from the roof or walls of a cavern, it appears to have originated in the multiplication of small squinch arches in an pendantive (hence the stalactite pendentive). It is a complicated arrangement of prisms and is a geometrical decoration stimulated by the Islamic prohibition of animal representation.[15] On the top side of the interior tomb a few signs of paint is still visible in lapis lazuli very similar to the colour used in Chand Minar and Chini Mahal inside the fort.

Second Maqbara :

Very close to the first Maqbara with similar tudor arches in all four sides is the second maqbara, built in brick and plastered. This structure also contains medallions in spandrels on either side of the top of the arch. It is more simpler than the first creation and its dome above is shaped like an inverted cup. It also lacks the burj feature of the first maqbara. The plaster is still visible but gradually it is falling apart (Plate LIV).

Unfortunately the maqbaras which once came under the protected monuments during the Nizam's dominion now is deprived of this protection. Neither is it included under A.S.I. which is generally responsible only for the national level monuments nor the State Archaeology which is responsible for the state level monuments. It lies forborn, isolated and neglected. Since it lies on the side of the main road continous vibrations will definitely destroy them. Also few structures are gradually a rising on its sides.

A Third Maqbara :

Is a little further down on the right side of the road and is much more simpler than the earlier elaborate maqbara. There are openings on all four sides as is the general feature of the maqbara. The arches are of the tudor type.

It is built in stone and not bricks. The upper portion of the chhajja is built in brick work as well as the pyramidal roof above. The dome above is very different and it is difficult to place it in any dynastic architectural context though the arches are similar to the tudor arches of the Tughlaq period (Plate LV).

Hazrat Bahauddin Ansari Urf Langot Band Ansari's Dargah :

Dargah of Saint Shah Bahauddin Ansari Shuttari Qadri's dargah. The tomb also is located on the way to Daulatabad. It is located a few steps away from the main road. The dargah is in the usual square shape with a dome above. Unfortunately the whole structure is recently painted in gaudy colours. The entrance is very .small and the worshipper has to stoop to enter. Ladies are not permitted entrance. In the centre of the interior block lies the tomb of the saint Bahauddin Ansari. According to the muezzin (whose forefathers lived years since decades) the great saint used to pray in the basement of the maqbara. In the corner of the room a narrow place with steps to descend, which leads to a small room in the right with enough space only for a single person. He has to literally crouch to enter the space where he used to withdraw for prayers and meditation. A little below is a open space with an opening which opens into a well through which the saint used to go to a secluded masjid within the well but on a higher plane.

The main dargah with the dome is also supported by squinched arches. Above on all corners of the dome are the usual burj or towers. The dome has an inverted lotus above topped with a kalash (Plate LVI).

Towards the right and opposite the dargah is a khankha which has been recently renovated as it was in ruins. The interior portion of the khankha is supported on pillars with arches.The walls too have blind arches.

Nagarkhan :

Is a part and parcel of the building complex within the enclosed wall of the dargah.The Nagarkhana is a masjid built by a disciple of Saint Bahauddin who was travelling by ship and the ship was caught in a storm. He offered prayers to the saint and promised to build a masjid in memory of the saint if he was saved from the danger. Fortunately he was saved and later he built a masjid and fulfilled his promise. The shape of the masjid is similar to a ship (jahaz). The upper portion is in the shape of a mast. On both sides are oriel window supported on brackets. Between the windows arched openings are built. There are several alterations in the original building, as several modern doors and windows are fixed on the ground floor. Part of the ground floor is used as an office.

A couple of years back the fifth centenary of the Hazrat was celebrated pompously by the dargah trustees.

Shahzadi Ka Maqbara :

This maqbara is in close proximity to the Shia Saints maqbara with the Balaghat hills as a backdrop is one of the most sublime and stately maqbaras. It is however difficult to identify the princess in whose memory it is built and lies buried securely. There were several Mughal princess (Plate LVII) during the time of Shahjahan as well as Aurangzeb who had a long stay in Aurangabad. Princess like Zinat un-nisa is mentioned who had her own palaces in Quile-Ark complex as well as Roshanara the sister of Aurangzeb is said to have accompained him continuously. All this is conjectural as no references have been found so far. It has grand entrance gateway with a three centtic arch. The arch is recessed with a small square doorway in its centre. On either side of the projecting gateway are carved pilasters and the whole scheme of the arch is carved amidst a cornice. The walls of the gateway which project outside and its surfaces are divided into a number of panels with diagrams each repetitive. Though the designs are same it appears neither monotonous nor stylized but lends it a beauty all its own. The main gateway is contained in between slender buttresses rising up gracefully.

On all four corners of the huge maqbara structure are the typical chhattris of Mughal period. The kiosks are places on a raised circular structure. The kiosks are pillared. The kiosks are surmounted by cupolas and pinnales.Its dimensions allow a number of commodious rooms to form the interiors, providing quarters for a guard.[16]

Here however it has a more decorative utility. There is a hollowed vault made in bricks. The cupola is completely fluted the arches are engrailed. From the interior of the maqbara all the kiosks can be reached by means of double staircases. In between the staircases on either side are rooms probably for residential purposes for guards.

The tomb structures were given a special place in the architectural creations during the early sultanate period which evolved into more complex nature during the Mughal period. Thus the grand compesition consists of majestic gates, pavilions, huge halls, spacious gardens all on grand scale yet enshrining a mere handful of dust. This tomb structure has a magnificent gateway, kiosks on all four sides, spacious gardens etc. But the main maqbara unlike the highly elaborate maqbara of the Mughal period is simple.

It is placed in the exact centre of the maqbara a raised platform of about three feet high and in the midst of this is the mortuary chamber with a grave or qabr. The tomb can be reached by four steps from all four sides (Plate LVIII).

On all four sides of the central maqbara are ablong reservoirs all better known as Char Bagh typical of Mughal structures. The four reservoirs once had active fountains and also fed water to the entire garden through internal pipelines. The Mughals laid great emphasis on gardens which was associated with architecture, and thus provided with various devices to provide adequate running water in and around the building. The Mughals built their edifices,chiefly their tombs in well planned and well laid out gardens with water channels, water chutes, tanks, fountains and other accessories lending it enhanced beauty.

On either side of the maingate chambers are built. The chamber on the right has a square ceiling, numerous square panels, embellished with decorative designs. On the back wall are engrailed arched designs with floral design tied in a knot shape. Within the engrailed arch is a mihrab.The engrailed arches are divided into five panels and two tiny panels in each corner. All the panels have carved designs in plaster of rounded designed bowls containing fruits like grapes, watermelon, custard apples and pomogrenate. On the left side walls are carved designs of wine bottles with rounded base and long narrow necks with a closed pointed knob, popularly called Suraahis (Plate LIX and LX).

The ceiling is ornamented with recessed square designs and below this are intricate carvings of star and diamond designs with tiny horse shoe shape arches bordered all around. The colours are still visible on its ceiling. The chamber on the left too has similar designs though it does not have the finish of the other chamber.

It is probably an interior quality of stucco plaster. The technique of stucco plaster is very ancient but during the Mughal period it was not used as background but as elaborate ornamentation in geometrical designs. Generally on the brick or stone surfaces a thick layer of plaster was found. Lime was first slaked and kept under water for a week or more, and was then mixed with double the quantity of sand, and the mixture than thorough ground. This plaster was applied to rough stone or brick walls in thin successive coatings, one after the other, each thoroughly beaten edgeways with thin strips of wood after wetting. The plaster finally became a quarter of an inch. The final coating on which the painting was done was prepared with marble lime, perfectly slaked and left under water for months, sometimes even a year for the best results. Curd was then mixed with it in proportion of half a seer to half a maund of dry lime (about 465 grans curd to 18/2 kilogramme of lime) The mixture is stirred well and allowed to stand overnight. Next day the water is strained and fresh water added. The process is repeated for a week when the lime is ready to be used.

The stucco was used for decorative designs and the red colour probably geru (red ochre) is used as a colour scheme for added effect.

Nizampur Sarai :

Sarai : Public buildings like sarais are the most conspicious of the public building like hammams, dak-chauki (post stations), kotwali (Police station), katra (market place), and kacheri (administrative office) It was introduced in the 13th Cent. by the Turks. The main features of the sarais may be listed thus.

1. Square or rectangular disposition enclosed on all four sides by masonary walls with entry through one or sometimes two gateways.
2. Series of rooms fronted by small vaulted spaces along all the four sides inside the enclosure. Warehouses in the corners of the enclosures.
3. Existence of a small mosque and one or more wells inside.[19]

During the Tughlaq period Firoz Tughlaq was a great builder of public works of public utility and is said to have built 200 carvan sarais.

Following the tradition of some rulers of the past Sher Shah lanted shade giving trees on both sides of the established road and sarais or rest houses at different stages with seperate arrangeement for Muslims and Hindus. These sarais also served the purpose of post houses, which facilitated quick exchange of news and supplanted the government with information from different parts of the empire.[20]

It was said that sher Shah constructed many roads in order to connect his capital with the various parts of his dominions. Four of his roads are well known. One which ran from Sonargaon in East Bengal through Agra, Delhi, Lahore to Indus was 1,500 kms in length. It was known as Sarak-e-Azam and might be identified with the modern Grand Trunk Road. On either side of the road were built 1,700 Caravan Sarais. Provision in the sara is was made to station horses for the servants of the postal department. Each sarai had a well and a mosque with an imam and a Muezzin. There was a police official (Shiqdar) posted at every sarai for keeping and preventing crime. Besides being useful for the travellers, the sarais were particularly meant to serve as rest houses for the servants and runners of the postal department who carried the royal mail. Some land in the vicinity of each sarai was endowed for its maintenance. In the words of Dr. Qanung, these sarais were veritable arteries of the empire and the road and sarais were essential for the success of administration.[21]

During Akbar's period too the various parts of the country and most of the important towns were connected by kacchha roads, shaded by trees on either side and clearly demarcated by sarais meant for merchants and travellers to spend the night in security and for the asylum of poor strangers.[22]

The Sarai at Nizampur: The Sarai is a distinctive structure, huge and spacious located within a small village of Nizampur bordered on its right by the village. The sarai is extensive and capacious.

Its expansive area is indeed impressive but presently it is in a dilapidated condition. The entrance once upon a time must be very imposing. It has a double arched gate with two gateways. The extreme ends of the corners of the gate are decorated with pilasters. It is a double storied gate with double openings on either side one above the other. On all four angles are burj. However the burj on the right corner has disappeared while the rest are intact. The burj is supported on an octagonal superstructures for effect.The kiosk or pavilions have a cupola roof supported on brackets and lintels. The cupola is lime plastered on the top of the cupola is the padmakosa (lotus petals) and kalasa finial. The pavilion has eight open engrailed arches. The Pavilion had open arches for the guards to watch from all sides. It served more as a watch tower. Between the main gate and the kiosks are brick walls with blind arches. Wile the main gateway and chhattis (kiosks) are made of stone blocks. The walls are all built in brick (Plate LXI).

In the interiors of the sarai are divided in 56 equal compartments and connected to each other by a system of arches arranged in rows with 14 compartments divided equally on all four sides with a spacious courtyard in the centre. The arches are mostly three centered. Each compartment is covered on the top by a domical vault of a simple but elegant design. In the centre of each side are Diwankhanas with huge open arches. These arches are engrailed. There is a well within the sarai. The right side Diwankhana was used as Ashurkhana, while the other two most have been used as police chowki.

The main entrance served as a secure gateway with double door system for the travellers. Sarais were meant both for the poor and rich. Between the two gateways is a raised platform with the interior octagonal in form with chambers for the guards. The gate is very similar to the Nawkhanda gate.

Unfortunately presently the beautiful sarai is a haven for nomadic people some of the residents have been staying since decades, while some make it their temporary home.

The sarai belongs to the Mughal period as is evident from a number of contemporary and distinctive architectural features of the Mughal times. The chhattis are a characteristic feature, the engrailed arches of the interior Diwankhana as well as the kiosks are typical of the Mughal times. There is a mosque and a temple. The mosque is located in front while the temple is located on the backside of the sarai. (Plate LXII).

Khuldabad: as stated earlier Khuldabad, the Abode of Eternity was also an important part of Daulatabad. The Location gained more importance during the Mughal period due to the special affinity of Aurangzeb for the heavenly abode. It also gained importa nce as a pilgrimage place due to the exodus of several sufi saints to Khuldabad in the 14th Century.

Its description as given by Claude Campbell in the Glimpses of the Nizam's Dominious is as follows, "Rauza or Khuldabad, lies five miles north of Daulatabad and fourteen from Aurangabad and is surrounded by a strong wall which was erected by Aurangzeb. It is about 2000 ft. above sea level, is pleasantly situated, has a temperate climate, and is largely used as a sanitorium during the place is 20ft in height. It has four gates, Nakar Khana, meaning a music room to welcome visitors, Pangra so called after a tank close by, Langru the eastern and largest gate and Mangalpait at the South of the Rauza as the name implies Kerbella of Deccan Mussalman of the Shia sect. It is celebrated as the burial place of many distinguished saints and royalty, amongst the royalty emperor Aurangzeb, his second son Azam-us-Shah, Asaf Jah the founder of the Hyderabad dynasty and Nazir Jung his second son, Malik Ambar, the powerful minister of the Nizam Shah of Ahmednagar, Tana Shah the exiled king of Golconda. Rauza is also famous for the ashes of the sufi saints Booran-ue-din and his brother, Zerbuksh besides many other holy men containing 15 to 20 domed temb and 1400 plain sepulchers.

According to Raunaq Ali the auther of Rauzat-ul Aqtab, this fort of Daulatabad was called Khuldabad—heavenly abode. It was known for the graves and tombs of Muslim Saints, kings, prince and princess.

Some of the important monuments in Khuldabad is the tomb of Aurangzeb. The tomb of Malik Ambar Bani Begum's Bagh.

The tomb of Aurangzeb : Originally the tomb consisted only of a wooden slab with a persian inscription. "No marble sheets should shield me from the sky as I lie there one with the earth.[25]

There is a legend connected with Aurangzeb's tomb well known for his diligence and piety, the emperor on his deathled decreed that only Rs. 14 and 12 annas should be spent on his grave. This was the sum he had earned himself by stitching caps. He had also earned Rs. 350 from Quranic inscription, but he said, "Don't use this money in case I've made a mistake in copying the Quran, as I will be answerable to Allah for that".

Since then the tomb has been embellished. In 1921 the Nizam of Hyderabad and Lord Curzon had it covered with marble and surrounded with a pierced marble screen. But a patch of earth with small sabza plant remains on top, it is covered with a plain white sheet, and it is roofed only by the vault of the sky.

Part of a large funerary complex Aurangzebs tomb has alongside it the graves of his son Azam Shah, and his wife, small, marble headstones within an enclosure of solid marble slabs.

A few steps ahead is the dargah of Saint Zain-ud-din, the last Chisti. Sect of Sufi Saints, who died in 1370. The mausolem has chase. silver doors and the threshold is embedded with curiously cut and polished stones. An embroidered bracade hall covers the grave itself, over which is suspended a string of ostrich eggs. Why ostrich eggs? Despite enquiry there was no logical answer and these probably. was no significance attached to its presence. Zian-ud-ddin was the religious priest of Aurangzeb hence when Aurangzeb died hes last wish was to be buried close to his spiritual teacher.

Directly opposite the Dargah of Hazrat Zian-ud-din lies the dargah of Hazrat Burhanuddin right across the road. He was the predecessor of Hazrat Zian-ud-ddin. Here were the tombs of Asaf Jah, the founder of Hyderabad dynasty and his wife. These are on a porphyry platform inlaid with white marble, with a high screen of the same red stone around them. Burhanuddin's dargah is as richly decorated as his successors. On the platform immediately in front one is shown little pellets of silver which seem to grow out of the stone. Again a legend is attached to the saint. It was said that after the saint's death in 1344, there was famine in Deccan. So his followers gathered before the dargah to ask him to intercede with the Almighty. To thenr astonishment, the nest morning they found silver plants growing out of the platform, of which todays silver pellets are the remains.

Saint Khwaja Hazrat Muntajebuddin Saab was another famous saint and the mureed(disciple) of Khwaja Nizam-ud-ddin Auliya of Delhi.Nizamuddin sent him to Deccan. Muhammad Qasim Ferishta the historian writes that several saints accompanied him to the Deccan and Khuldabad.[26] But the main problem was to run the livelihood of so many saints. A miracle is associated with Hazrat Muntajebuddin that every night a small piece of gold used to come out from the ground while some said every night he got a box of gold which he sold and sustained the people. The reason later he came to be called as Hazrat Khwaja Muntaz-bud-din Zar Zari Zar Baksh Dula Miya. Also a Chisti saint, he was the khalifah or most mature disciple of the Auliya.

Malik Ambar's tomb is also located at Khuldabad. He was the prime minister of Nizam Shahi dynasty and the commander who played a dominating role in political rule of Daulatabad.[27] During his stay at Daulatabad he is said to have constructed his own mausolem in stone. Constructed on a high platform the entire construction is in black stone. A beautiful domed square building with kiosks on all four openings with broad lintels above. The dome has lotus petal encircled as a border in its lower portion, above is the usual kalasha. The lower portion is embelished with large jalis or screen. The upper portion is bordered by arched motifs.

To the left the Malik Ambar's mausolem is a guest house for travellers with kitchen, hamam and toilet facilities of the same period. A little distance away from Malik Ambars tomb is his wife's tomb whose name was Siddi Karuna also made in stone. A little distance away from here is a mausoleum of Nizam-ul-Mulk Baheri.

Among other rulers the grave of Abul Hasan Tana Shah also nearly. He was the son-in-law of Abdullah Qutub Shah and the last ruler of the Qutb Shahi dynasty. Other important saints who rest here are Hazrat Momin Arif, Amir Hasan Delhvi a close friend of the famout poet Amir Khusrao. The burial places of these two are however far away from Khuldabad.

Bani Begum's Bagh : The original name was Zahan Banu Begum, the grand daughter in law of Aurangzeb whose tomb lies within a well decorated garden. It has a well designed in the shape of a square tank. The entire pathways are made in red stone. The tomb is laid with red stone. The entire maqbara was made in marble and red stone. Unfortunately numerous people removed the marble and left bold patches behind.

The Fairy tale tank : The Pariyon ka-talab as it is popularly known, is situated two kilometres south af Khuldabad is more than just a scenic spot. One of the lesser known places of tourist interest in this region, is a large man-made water tank. The custom is to take a bath for women in the water of the Pariyon ka talab and the water is recommended to cure ailments.

The tank dates back to the 12th Cent.A.D. and is associated with Jalal-ud-din Ganj-e-Rawa who was a contemporary of the better known Shamshuddin Iltutmish, the slave who went to become the Sultan of Delhi. Jalal-ud-din came down to Khuldabad, he captured a jinn(Mythical demon) named Azar and kept him bound and imprisoned under the stone, on which he sat and prayed. To this day according to legend there is a stone on the east of Jalal-ud-din's tomb which is thought to be the same under which the demon was incarcerated.

The Hazrat power to heal and his ability to bestow fertility through the fruits and waters of the pariyon ka talab finds continuity in the many legends and folk tales about other saints of Maharashtra.

The entire region of Daulatabad and Khuldabad is steeped in history and consists of several secrets. There are also several legends one of which says that Malik Ambar had dug a secret path underground from Aurangabad to Daulatabad. Since so many dynasties ruled, transferred their capitals it is no wonder that the entire region should be spread with so many historical remains some of which are in a very good state of preservation while some are literally in ruins. Unfortunately most of the monuments around the fort lie neglected as they have no caretakers.

Due to financial crunch no government agency is prepared to take responsibility of restoring these ancient and medieval structures. Unless immediate steps are taken most of these structures around the fort will collapse. Some of the structures have come under private ownership. If the owners are interested they spend a lot in the restoration of the structures if not they lie neglected.

REFERENCES

1. Battuta Ibn.: Travels in Asia and Africa (1325-54) Published by George Routledge & Sons, London 1929, p.227.
2. Aurangabad District Gazetteer, 1977, p. 966.
3. Raunaq Ali- Rauzat-ul-Aqtab, p. 22
4. Brown Percy. Indian Architecture Islamic Period, p. 3.
5. Ibid., p. 3
6. Ibid., p. 3
7. Ibid., p. 16
8. Aurangabad District Gazetteer, p. 87.
9. Broun Percy. Indian Architecture, p. 21
10. Aurangabad District Gazetteer, p. 127.
11. Course in Tourism IGNOU, Tourism. The Cultural Heritage T5-1, Block 7, 1994, p.19.
12. Brown Percy, p. 15.
13. IGNOU. Tourism. The Cultural Heritage, p. 19.
14. Srivastava. Ashirbadi Lal. Medieval Indian culture. Agra, 1964, p. 150.
15. Nath R., p. 73.
16. Brown Percy, p. 100
17. Nath R.Colour Decoration, p. 2
18. Shukla Vastusastra Vol. II, p. 393.
19. IGNOU. Block 7, p. 22.
20. Majumdar & Raychaudhari & Dutta—An Advanced History of India , p. 441.
21. Srivastava. The Mughal Empire, p. 111
22. Ibid. Pg .112
23. Aurangabad District Gazetteer, p. 390
24. Raunaq Ali- Rauzat-ul-Aqtab, 1994, p. 25.
25. The Taj Magazine Vol.23, No.2 Second Quarterly 1994, p. 25.
26. Raunaq Ali, p. 46 to 48.
27. Ibid., p. 68.

CHAPTER - VI

CONCLUSION

The quality of building environment around most Indian monuments is steadily deteriorating and this is so true of Daulatabad fort so rich in medieval architectural structures towering minar commanding attention, colossal fortification walls surrounding comprehensive area elevating its towering presence. However presently it is pitted against ugly infrastructure, unplanned and illegal constructions and decaying civic institutions. Even as the Shining India' slogans, advertisements. IT superiority are being publicized, economy liberalized, urban governance appears to be stuck in the ancient regime. The causes are well known; rampant encroachments, defunct civic bodies, corrupt officials, political support, unimaginative or no planning. This general laxity has enervated our entire system.

While on the other hand rapid economic growth has transformed countries like Malaysia, Singapore, Dubai with gleaming shopping malls, spacious, high rising, high quality seven star hotels have changed these cities beyond recognition. All this is in sharp contrast with the unimaginative concrete blocks superimposing our heritage wealth.

However there are improvements in the World Heritage Monuments surroundings. The outskirts of Ajanta and the general surroundings of Ajanta have changed beyond recognition with high level consultancies like the Tata consultancy Ajanta has been developed with lawns, gardens, water chutes, well planned shops, bridges, parking area, landscaping around Ajanta caves proper.

But the same is not true of Daulatabad, with shabby restaurants opposite the fort, unplanned parking areas, fruit carts creating garbage around, cheap souvenir shops and a narrow road in between. The road between the shops and the fort is so narrow that several accidents have killed/precious life. As it is an highway the traffic flow is very speedy. Thousands of tourists cross the road after returning from the fort to drink as well as eat in the way side restaurant.

There have been several efforts by the officials of Maharashtra Tourism Development Corporation to rearrange these dhabas by pushing them back to increase space between the shops and the fort. M.T.D.C. had even won a case against the shopkeepers for relocating them but political support completely unsettled their plans. As the Ajanta³Ellora Development plan has also included a huge amount for the beautification around Daulatabad fort, planned development of the area by creating gardens, lawns, a well defined parking space, the possibility of transforming the ugly surroundings had increased. Unfortunately as the land belongs to Wakf Board the situation is difficult. The plans also include construction of shops suitable to the surroundings. M.T.D.C. is prepared to rent shops, restaurant to all the present owners but to no avail.

In the second phase of the development plans the Forest Department has decided to co-operate with M.T.D.C. The land behind these shops falls under the forest department. The plan is to beautify it with an aesthetically planned garden area, an architecturally suitable cafeteria and toilet blocks.

One of the major problem around all A.S.I. owned monuments are lack of toilet facilities. Tourists spend hours ascending and descending the fort and natures call has to be controlled. A.S.I. officials claim that they are custodians of the monument hence their priority is the maintenance of the monuments and not providing facilities to the tourists.

Since the Japanese Bank for International Co-operation is funding the beautification around Daulatabad fort this is a rare opportunity to create a special , character' that will live beyond current architectural fashion. Sheer indifference of ASI might allow gradual decay of the huge massive fortification walls due to close proximity of parking space.

In this 21st century preserving such antiquated monument is an absolute necessity. We have to be especially conscious about this issue because the transformation process is very fast. Due to growth and economic prosperity alternatives for the land will begin to tempt developers who might replace these small shops and dhabas with high rising buildings and shopping malls in the future.

The way to saving the surroundings from such a situation is to convince the owners and the Wakf Board and to find imaginative ways of re-building these shops which architecturally match the forts architecture. The current conservation strategy, wherever it exists especially ASI norms and rules is merely to ban all change. This is counter productive as it dooms architecture to steady degeneration and irrelevance.

Daulatabad surroundings require urgent face-lift and the participation of NGO's is also necessary to push through the project. Some noted historians like Dr. R.S. Morwanchikar are instrumental in attracting the attention of the state government and UNESCO to the inclusion of the fort as a World Heritage site. UNESCO according to latest information is in the process of identifying Daulatabad as a World Heritage site, which is a major success story.

Another important additional attraction which is in the pipeline since a decade is the sound and light show inside the fort. There is ample prime space for the show and a very fascinating history but due to lack of finance general apathy of all concerned to give a concrete shape to the idea the scheme was never implemented. The script, sound system and the most modern technical equipment is ready with private agencies, but it is still a flash in the pan.

BIBLIOGRAPHY

1. Agarwala V.S.	"Indian Art" (Varanasi, 1965)
2. Bashiruddin Ahmed	"Waqiat-e-Mamlikat-e-Bijapurr" Agra, 1915
3. Batuta Ibn	Rehla—Translation, Mehdi Hussain, Baroda, 1953
4. Batuta Ibn	"Travels in Asia and Africa", Translation, H.R. Gibbs London 1923
5. Bernier F.	"Travels in Mughal Empire" (Ed. V.A. Smith Oxford 1914)
5. Bhandarkar R.G.	"Early History of the Deccan" Publisher Sushil Gupta Calcutta, 1957
7. Bilgrami S.A.	"Landmarks of the Deccan, AsifJahi", Aurangabad-1922
8. Bilgrami S.A.	"Historical and Descriptive sketches of H.H. Nizam's Dominion" Vol. II Times of India Steam Press, Bombay 1884
9. Bilgrami S.A.	"Rauzatul Auliya" Urdu Translation. Mohammed Abdul Majeed Karni Bazar, Hyderabad 1926
10. Brown Percy	"Indian Architecture Islamic Period", Bombay 1942
11. Burgess James	"The Muhammadan Architecture of Ahmedabad" Part II (ASI New Imperial Services) Vol. XXXII London 1903
12. Campbell	"Glimpses of Nizam's Dominion" 1898 H.H.
13. Chawla Rupika	"Restoration needs Restoration" Loktnat Times January 10,2000
14. Cousens K.	"Guide to Bijapur" S.M. Industries Co., Poona, 1905
15. Cousens	"Bijapur and its Architectural Remains" K.B. Publication, New Delhi, 1977

16. Geswell — "Early Muslim Architecture" Vol.I Oxford, 1932

17. Derck Hill and Oleg — "Islamic Architecture" Brabar

18. Dikshit K.N. — "Towards the Formulation of National Policy of Conservation of Monuments"

19. Dutt B.B. — "Town Planning in Ancient India"

20. Fazl Abul — "Ain-e-Akbari Blocklunan" Delhi, 1965

21. Ferishta M.K. — "History of the Rise of the Muhammadan Power in India" Vol. I English Translation Briggs John, Calcutta 1966

22. Ferishta M.K. — "Gulshan-e-Ibrahim-Luchuois Tarikh-e-Ferishta", Kanpur, 1884

23. Fergusson James — "Architecture of Ahmedabad", London, 1960

24. Fergusson James — "History of Indian Architecture" London, 1910

25. Fergusson James — "Indian and Eastern Architecture", London, 1876

26. Fletcher Sir Banistor — "A History of Architecture On Comparative Method", London, 1954

27. Ghaussuddin Syed — "Turkish Bath" (Urdu) Hyderabad, 1975

28. Godard Andre — "The Art of Iran" (Michael Reran) London 1965

29. Hai Abdul — Gulistan-e-Khuldabad

30. Raig W. — "An Inscription in Daulatabad Fort" (Epigraphy Indica) 1907-8

31. Raq abdul — "Akhbarul Akhyar-to-Asrarul Abrar", Delhi 1333 AD.

32. lsan Maulana — "Futuhus Salatin Tram Mehandi Hasan", Agra 1938

33. Iqtidar Russain and Qazi Muhamad — "Shihabuddin At Umaris Masalik At Absar F. Malak At Ahmed Ansari" (Atigarh 1971)

34. Jairaj Bhoj — "Art and Cities of Islam" New York, 1964

35. Jairaj Bhoj — "An Outline of Islamic Architecture" New York 1912

36. Joshi Lakshman — "The Deccan Forts Bombay" 1961 Kamlapurkar

37. Kamra Krishan — "Managing Tourist Destination" Kanishka Publication, New Delhi

38. Khusru Amir — "Khazainal Futuh" Trans Muhamed Rabib, Bombay.

39. Kunte B.G. — "Ahmednagar Chi Nizam Shahi", Bombay, 1962 (Marathi)

40. Kunte B.G. — "Maharashtra State Gazeteer History" Part II Bombay 1972

41. Kunte B.G. — "Aurangabad District Gazetteer" Bombay 1977

42. Lahori A.H. — "Badshah Nama" Vol.I Bibliotheca India Collection of Oriental Work Royal Asiatic Society of Bengal Calcutta.

43. Majumdar, Ray — Macmillan and Co. London 1948 Chandhari and Dutta

44. Manucci — Vol.II Storia Do Mogar Trails, Willian llvine, London, 1907

45. Molu Noorul Hasan Nayyar — "Noor-ul-Lughat" Vol.II Lucknow, 1927

46. Morwanchikar R.S. — "Deogiri—Daulatabad An Archaeological Overview" Publisher, Rajwade Sanshodhan Mandal Dhule 1993

47. Mulk Raj Anand — Marg No.I XXIV December, 1970

48. Nath R. — "History of Decoration Art in Mughal Architecture", Delhi, 1976

49. Nath R — "Some Aspects of Mughal Architecture" New Delhi, 1976

50. Nath R. — "Colour Decoration in Mughal Architecture"

51. Narwane M.S. — D.B. Taraporevala and Sons Bombay 1970 Forts of Maharashtra Aptt Publisher Cor, New Delhi, 1995

52. Pagadi Madhavrao — "Studies in Maratha" Vol. II Shivaji University Publications, Kolhapur, 1971

53. Pigman Augustus — "A Story of Water" New York, 1938

54. Radhakrishnan Dr. — "The Cultural Heritage of India" Vol. I (Calcutta Institute of Culture), 1970

55. Rafi Ahmed Abu — "Studies in the History of Medieval Deccan" Adarah Adbiyat-e-Urdu, Delhi, 1977

56 Raunaq Ali — "Yadgar-e-Daulatabad Khuldabad" 1310 A.D.

57. Raunaq Ali — "Rauzatul Aqtab" Aurangabad, 1339 A.D.

58. Riyaz Ansari — "Medieval Daulatabad Complex A Cultural Study" Jan. 1983 Aurangabad (Thesis)

59. Rushbrook William — "History of India"

60. Safiuddin Siddiqui — "Shaikh Ruknuddin and his Role in Medieval Deccan". Islamic Culture, 1981

61. Seely John B. — "The Wonders of Ellora" II Edition London, 1825

62. Seshadri A.K. — "Conservation of Monuments" in India Book Publishing Co. Delhi, 1997

63. Shahnawaz Khan — "Maseer-ul-Umera-Persian Manuscript" Copied by Ram Raje in 1770 Hyderabad

64. Shrivastava Ashirvadi Lal — "Medieval Indian Culture"

65. Shukla D.M. — "Vastu Shastra" Lucknow, 1967

66. Siddiqui Mohammed Suleman — "Sufis of the Deccan" Hyderabad, 1975

67. -do- — "Chisti Order in Deccan Islamic Culture", July 1978

68. Sinha and Banerjee — "History of India" Calcutta, 1955

69. Smith V.A. — "Oxford History of India" Oxford University, 1958

70. Stuart Piggot — "Pre-Historic India to 1000 B.C." Penguin Press, 1961

71. Swamy Kumar Anand — "Early Indian Architecture Cities and City Gates", Journal of Eastern Art Vol.2

72. Sylvia Crown and — "The Garden of Mughal India" (London 1929) Sheila Heywood

73. Taverair Jean Baptiste — "Travels in India" Ed. William Crooke, London, 1925

74. Thevenott Careri — "Indian Travels of Thevenot and careri" Ed. Surendranath Sen, New Delhi, 1949

75. Thosar M.S. — "Studies in the Historical and Cultural Geography of Marathwada" Thesis submitted to Marathwada University

76. Toy Sidney — "The Stronghold of India" London, 1957

77. -do- — "Castles a History of Greek Fortification"

78. Useino — "Monuments of Azerbaijani Architectures"

79. Verma O.P. — "The Yadavas And their Times" Nagpur, 1970

80. Verma Amrit — "Forts' of India", Publication Division Ministry of Information and Broadcasting, Govt. of India, May 1985

81. Winter F.E.	"Greek Fortification", (London, 1970)
82. Yahya Sirhind	"Tarik Mubarakshah", Elliot and Dowson
83. Yazdani Ghulam	"'Bidar', Its History and Monuments", (London, 1947)
84. Yazdani Ghulam	"The Early History of the Deccan", Vol.I and II London, 1960 Other Publications
85. Gazetter of Aurangabad	Published under the Order of H.H. Nizam, Govt. Bombay, 1884
86. Aurangabad District Gazetteer	Bombay, 1977
87. Reports of Archaeological Department	1929-30
88. Encyclopedia Britannica	Vol.3, 1768
89. Bharat Ithas Sanshodhak Mandal—Quarterly	July 1935
90. Jame-e-Kamil Suba	Jat-e-Deccan-Persian Manuscript Salarjung Museum Library, Hyderabad
91. Journal of Society of Architectural Historian	XVII No.3, 1958

INDEX

***

PLATES

Plate I. Ambar Kot

Plate II. Partial remains of teakwood beam

Plate III. Entrance Gateway

Plate IV. Strong Bastions and battle mented parapet

Plate V. Chambers with pillars

Plate VI. Lawns and fortification walls with bastion and hanging window.

Plate VII. Semi-circular tower, Naggar Khana

Plate VIII. Guard rooms

Plate IX. Hathi tank

Plate X. Storage for ammunition

Plate XI. Gateway

Plate XII. Bridge (Interior)

Plate XIII. Moat (Interior)

Plate XIV. Entrance to the Dark Passage

Plate XV. Bastions and fortification wall

Plate XVI. Mendha Top (canon)

Plate XVII. Passage through Gateway

Plate XVIII. An Ariel view of the fortification defence walls.

Plate XIX. Jama Majsid cum Bharat Mata Mandir Entrance Gateway

Plate XX. Ornamented Pillars of Bharat Mata Mandir

Plate XXI. Jama Masjid/Bharat Mata Mandir

Plate XXII. Interior Courtyard & Pillars of Jama Masjid

Plate XXIII. Step well at Daulatabad, Saraswati Step well.

Plate XXIV. Hammam (Vaulted roof)

Plate XXV. Water Channels inside Hammam

Plate XXVI. Diwan-e-Aam or Amkhas Palace

Plate XXVII. Pavilions also know as Diwan-e-Aam

Plate XXVIII. (A) Chini Mahal where Abul Hasan Tana Shah was imprisoned with Glazed tiles

Plate XXVIII. (B) Chini Palace (Sultan Abul Hasan Tana Shah)

Plate XXIX. Sonheri Mahal (Entrance Gateway)

Plate XXX. Sonheri Mahal (Inner Chambers)

Plate XXXI. Sonheri Mahal (Wood Work)

Plate XXXII. Remains of Rang Mahal

Plate XXXIII. Shahajahan's Baradari Mahal

Plate XXXIV. Interior portion of Baradari Mahal

Plate XXXV.Chand Minar (close up)

Plate XXXVI. Chand Minar from a distance

Plate XXXVII. Chand Minar (brackets)

Plate XXXVIII. Blue glazed tiles on Chand Minar

Plate XXXIX. View of Second Minar

Plate XL. (A) Brick kilns emitting smoke behind the fort

Plate XL. (B) Rows of Bricks

Plate XLI. Masjid (Grass and trees growing around it)

Plate XLII. Chini Mahal (verge of crumbling)

Plate XLIII. Rang Mahal (in ruins)

Plate XLIV. Sonheri Mahal (in ruins)

Plate XLV. Eidgaah-Khilji (Front Portion)

Ptate XLVI. Eidgaah-Khilji (Back Portion)

Plate XLVII. Jam-e-Masjid

Plate XLVIII. A secret Masjid inside the well

Plate XLIX. Shikargaah-Hauz-e-Qutlaq (side portion)

Plate L. Shikargaah-Hauz-e-Qutlaq (back portion)

Plate LI. Hathi Khana near Shahi Musafir Khana

Plate LII. A Dargah of a Shia Saint near the fort

Plate LIII. An empty Mausoleum near the fort (First Maqbara)

Plate LIV. Remains of Mausoleums and tombs near the fort (Second Maqbara)

Plate LV. An empty Mausoleum near the fort (Third Maqbara)

Plate LVI. A part of Dargah Hazrat Bahauddin Langot Band Ansari

Plate LVII. Mausoleum of a Mughal Princess (interior)

Plate LVIII. Interior of Mausoleum (decoroative motif)

Plate LIX. Wall decoration of Suraahis inside the Mausoleum

Plate LX. Nizampur ki Sarai (the Entrance gate)

Plate LXI. Nizampur ki Sarai (the interior portion)

Plate LXII. Nizampur ki Sarai (a view)

Plate LXIII. A View of Daulatabad Fort Environment

Plate LXIV. A View of Daulatabad Fort.